AF707807

Nanoelectromechanical Systems

Series Editor
Robert Baptist

Nanoelectromechanical Systems

Laurent Duraffourg
Julien Arcamone

iSTE

WILEY

First published 2015 in Great Britain and the United States by ISTE Ltd and John Wiley & Sons, Inc.

Apart from any fair dealing for the purposes of research or private study, or criticism or review, as permitted under the Copyright, Designs and Patents Act 1988, this publication may only be reproduced, stored or transmitted, in any form or by any means, with the prior permission in writing of the publishers, or in the case of reprographic reproduction in accordance with the terms and licenses issued by the CLA. Enquiries concerning reproduction outside these terms should be sent to the publishers at the undermentioned address:

ISTE Ltd
27-37 St George's Road
London SW19 4EU
UK

www.iste.co.uk

John Wiley & Sons, Inc.
111 River Street
Hoboken, NJ 07030
USA

www.wiley.com

© ISTE Ltd 2015
The rights of Laurent Duraffourg and Julien Arcamone to be identified as the authors of this work have been asserted by them in accordance with the Copyright, Designs and Patents Act 1988.

Library of Congress Control Number: 2015936236

British Library Cataloguing-in-Publication Data
A CIP record for this book is available from the British Library
ISBN 978-1-84821-669-3

Contents

Preface

Since the invention of the first calculating machine by Blaise Pascal in 1642, mechanical systems have contributed a great deal to the industrial revolution and continue to play a fundamental role in our daily lives.

Figure 1. *Pascaline (source: IBM)*

In the 1980s, mechanical systems moved to the micrometric scale and became micromechanical systems. Their lateral size ranges from several microns to several hundreds of microns for a thickness of 10 µm. These are transducers and their uniqueness lies in their ability to transform mechanical energy (movement, constraint) into electrical energy. The most well-known transducers are micro-accelerometers, gyrometers and pressure sensors; they have countless applications that are used by the general public (airbags, mobile phones, games, etc.).

Figure 2. *Micromotor (source: MEMSX)*

Several years ago, mechanics combined with electronics became nanometric. Nanosystems therefore infiltrated the world of mesoscopic physics, working at the molecular or supramolecular scale at sizes varying from 1 nm to 1 µm. These objects are the latest probes capable of measuring molecular interactions in physics, chemistry or biophysics. They cover a large number of applications, ranging from signal treatment to ultra-weak stimuli detection. Specifically, their low masses (10^{-15} to 10^{-18g}) make them ideal candidates for identifying macromolecules in the living world or measuring the cell strength. The potential of these components suggests that they will play a major role in medical diagnosis, environmental monitoring and food quality monitoring. This book will present the theoretical and technological elements of nanosystems. I hope it will prove to be a useful tool for future readers and will provide a picture – although probably incomplete – of current and future research in the field.

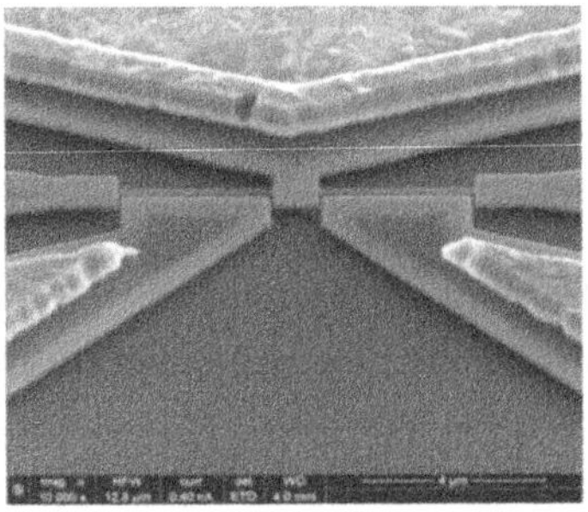

Figure 3. *Silicon nanowires (source: CEA-LETI)*

Laurent DURAFFOURG
April 2015

Physical Constants

ε_0: vacuum permittivity (8.85×10^{-12} F)

μ_0: vacuum permeability ($4\pi \times 10^{-7}$ H.m^{-1})

h: Planck constant (6.62×10^{-34} J.s)

$\hbar$: reduced Planck constant ($h/2\pi$)

k_B: Boltzmann constant (1.38×10^{-23} J.K^{-1})

m_e: electron mass (9.105×10^{-31} kg)

Da: 1 Dalton–atomic mass unit (1.6605×10^{-24} kg)

e: electron charge (1.602×10^{-19} C)

N_A: Avogadro number (6.022×10^{23})

c: speed of light (2.997×10^{8} m.s^{-1})

Notations

x, y, z: displacement (m)

v_x, v_y, v_z: speeds ($m.s^{-1}$)

a_x, a_y, a_z: accelerations ($m.s^{-2}$)

m: total NEMS mass (kg)

m_{eff}: effective NEMS mass (kg)

k: stiffness of NEMS ($N.m^{-1}$)

k_b: Boltzmann constant ($J.K^{-1}$)

k_{eff}: effective stiffness of NEMS ($N.m^{-1}$)

k_d: Duffing stiffness ($N.m^{-3}$)

l: length (m)

w: width (m)

t: thickness (m)

g: electrostatic gap (m)

E: Young's module (P)

I: moment of inertia (m^4)

σ: constraint (P)

ε: elongation

ρ: density ($kg.m^{-3}$)

T: temperature (K)

ℓ: mean free path (m)

q: charge (C)

C_{th}: thermal conductance ($J.K^{-1}$)

G_{th}: thermal conductance ($W.K^{-1}$)

R_{th}: thermal resistance ($W^{-1}.K$)

σ_{th}: thermal conductivity ($W.K^{-1}.m^{-1}$)

N_a, N_d: level of acceptor (P), donor (N) dopants (cm^{-3})

R: electrical resistance (Ω)

C: electrical capacitance (F)

μ_e: electrical mobility ($m^2.V^{-1}s^{-1}$)

ρ_e: electrical resistivity (ohm.m)

$\Delta R/R$: relative resistance variation

P_{th}: dissipated thermoelastic power (W)

P_e: dissipated electrical power – Joule effect (W)

G, γ_G: piezoresistive gauge factor

Q: quality factor

f_1: resonance frequency of the first mechanical mode of NEMS (Hz)

f_n: resonance frequency of the nth mechanical mode of NEMS (Hz)

δf: frequency dispersion around the nominal frequency (Hz)

Δf: increase of frequency in relation to nominal frequency (Hz)

ω_1: angular frequency ($2\pi f_1$)

ω_n: angular frequency ($2\pi f_n$)

1

From MEMS to NEMS

1.1. Micro- and nanoelectromechanical systems: an overview

This chapter will begin with a definition of nanoelectromechanical systems (NEMS), based on microelectromechanical systems (MEMS): at least two of the latter's three geometric dimensions were drastically reduced to nanometric size. Microsystems are miniature electromechanical transducer components used in actuators or sensors [BUS 98, MAD 11]. They contain a mechanical element that can move (Figure 1.1) under the action of a force. This force can be caused by the physical stimulus to be measured, such as pressure difference (Figure 1.2) or acceleration (Figure 1.3). The actuation force can also be artificially induced when the mobile body needs to be placed in controlled movement. In Figure 1.4, a membrane is actuated by an electrostatic force and constitutes a radio frequency (RF) microswitch. Figure 1.5 presents the different types of MEMS such as an optical micromirror, a microcantilever (microbeam) and a vibrating plate, which can be used as an RF clock.

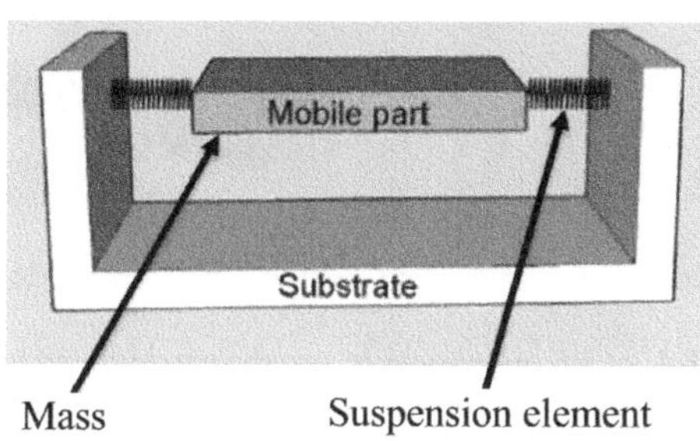

Figure 1.1. *Representation of a suspended mechanical structure containing a mass attached to the support by two suspensions enabling it to move (laterally and/or horizontally)*

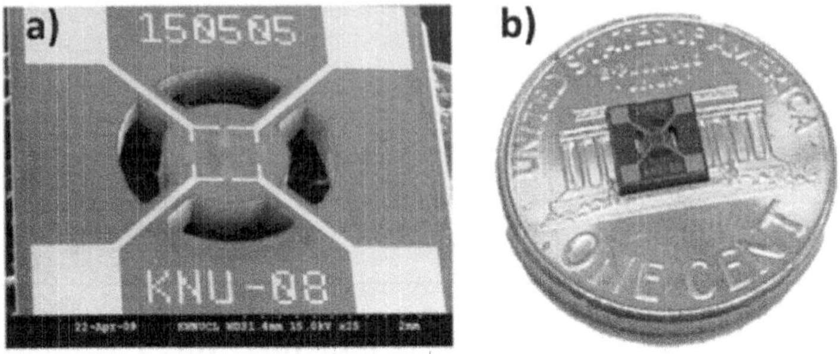

Figure 1.2. *Microsystem with a suspended membrane whose movement is caused by the force to be measured: a) micro pressure sensor (vertical movement of the membrane as a result of a pressure gradient); b) size of this component compared with a one cent coin [KIM 12]*

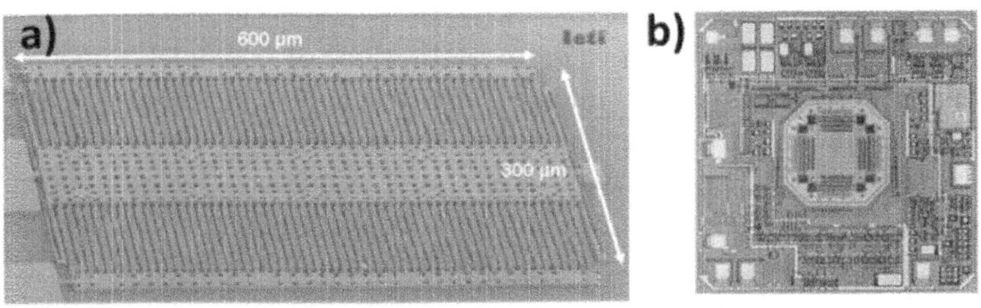

Figure 1.3. *a) Microaccelerometer with capacitive detector via interdigitated electrostatic combs (horizontal movement of the test mass subjected to acceleration) (source: LETI); b) example of an integrated circuit serving this sensor (source: ST Microelectronics)*

Figure 1.4. *a) RF microswitch: membrane actuated vertically by an electrostatic force enabling or not enabling ohmic contact between the two tracks depending on its position (source: LETI); b) the packaging of this component and the component on a circuit*

In development since the 1970s, MEMS have been used in aeronautics [BAR 11], automobiles and consumer applications since the 1990s. For instance, inertial sensors such as microaccelerometers (see Figure 1.3), gyrometers or magnetometers have been integrated on a large scale into cars (acceleration sensor for the release of the airbag), games consoles (Wii) and even mobile phones [YAZ 98, BEL 05]. Microactuators are used in video-projectors, microphones or to constitute micropumps for medical fluidic applications [THI 00]. More recently, MEMS have started to be integrated into RF circuits as switches (Figure 1.4) or reference clocks (Figure 1.5) [NGU 00]. MEMS have now reached a sufficient level of maturity such as to leave the field of pure research and are currently being developed directly in research and development (R&D) by big industrial players such as Bosch and ST Microelectronics [YOL 12].

Figure 1.5. *a) Analog devices two axes micromirror: the plate can move around two axes. These torsional movements are caused by an electrostatic moment [AKS 03]; b) a cantilever actuated horizontally by an electrostatic force between the beam and the control electrode [MIL 10]; c) square vibrating membrane forming an RF oscillator actuated by capacitive force [ARC 10]*

NEMS appeared in the early 2000s. Unlike MEMS, NEMS are an emerging technology. NEMS have long been tools devoted to fundamental studies to probe mesoscopic physical mechanisms. Their extremely small size makes them extremely sensitive to any external stimulus. To demonstrate this, Figure 1.6 shows the typical size of several mechanical structures making the move from microaccelerometers – as present in car airbags – to suspended silicon nanowires whose electrical and thermal conducting properties are profoundly modified by the effect of size.

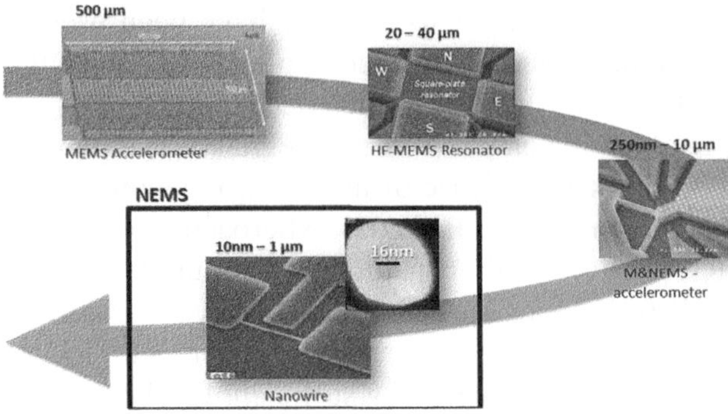

Figure 1.6. *From MEMS to NEMS: typical sizes*

Their extreme sensitivity opens the way to entire sectors of applications in biochemical analyses, which were not addressed by their big brothers, MEMS [EOM 11], such as force sensors [ARL 06, RUG 04] and ultra-sensitive mass sensors [EKI 04a, YAN 06, CHA 12] (shown in Figure 1.7). The principle of measuring mass landing on the surface of NEMS is relatively simple: it consists of monitoring the frequency shift of a NEMS kept at vibration (at a fixed controlled amplitude) using a closed loop circuit via a "Phase Locked Loop" (PLL) or an auto-oscillating circuit. Any element left on the surface of the nanosensor induces a frequency change, which we measure continuously (the mass and/or surface constraints change). This is known as a gravimetric effect (see Figure 1.8). Naturally, this detection principle can be combined with one or more principles, such as the variation of the conductance of NEMS caused by a variation of the electrostatic surface potential to potentially increase the number of measuring parameters and thereby precision. Pushing this approach forward, NEMS may be used as core components of a future mass spectrometer for proteomic analyses on a cell by cell basis [NAI 09, HAN 12]. In the short-term, NEMS will probably be used as sensors in multi-gas analysis systems. Details about these two applications will be provided in Chapter 5. Over time, this technology will be able to quantify gases that are potentially dangerous such as volatile organic compounds (VOCs) to analyze indoor air quality and even identify biomarkers in breath [ARC 1, BAR 12a, FAN 11]. "NEMS" technology can also be used to constitute new probes for near-field tools (magnetic resonance force microscopy and atomic force microscopy) [MID 00].

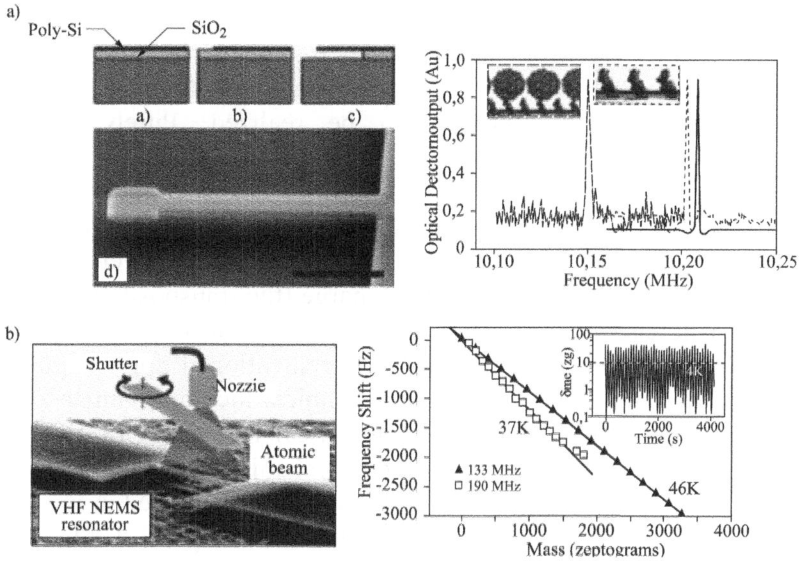

Figure 1.7. *Example of mass measurement: a) Baculovirus measurements according to Ilic et al. [ILI 04]; b) xenon atom measurements according to Yang et al. [YAN 06] (the noise floor is 7 zg or 30 xenon atoms). For a color version of the figure, see www.iste.co.uk/duraffourg/nems.zip*

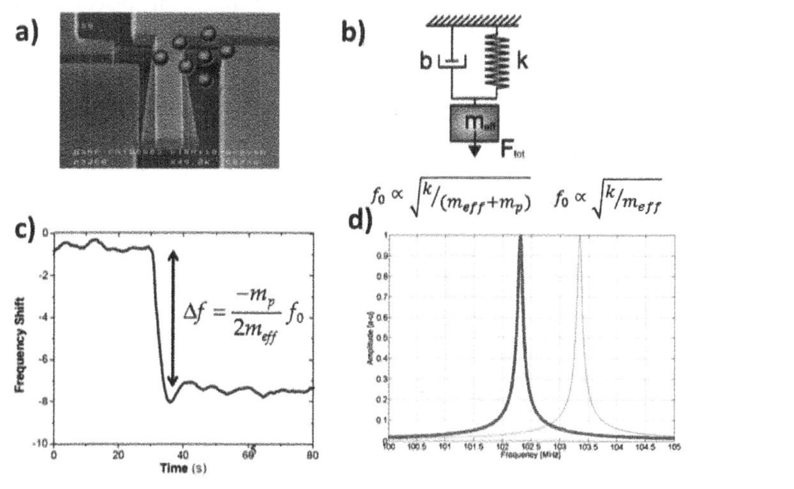

Figure 1.8. *The principle of mass measurement: a) example of a nanoresonator on which particles are left (nanocantilever); b) equivalent 1D model of a mass/spring-type resonator; c) shift in frequency caused by the arrival of a particle. The frequency jump allows us to deduce the accreted mass. Running the frequency brings the frequency back up to what is left on the mass; d) from the spectral perspective: shift in the spectrum toward low frequencies*

Mechanical nanostructures have by and large moved beyond the applicative field of sensors and can be used as switches/nanorelays in microelectronics. More specifically, nanoswitches feeding low circuit consumption (or power gating) can be realized. Purely mechanical components have little or no leakage in comparison with classic metal-oxide semiconductor field-effect transistor (MOSFET)-transistors. They can also replace metal-oxide semiconductor (MOS) transistors in direct current (DC/DC) converters, enabling the energy available from a given source to be transposed into a useable form from its charge (the transposition takes place by grinding the energy of the source with the switches). In addition to the replacement of MOS transistors in the aforementioned applications, some researchers have imagined reaching a mechanical memory similar to that of Charles Babbage's 19th Century mechanical calculator. As a result, a great number of works on nanorelays based on mechanical elements, as shown in Figure 1.9, or on NEMS/transistor hybrid elements, displayed in Figure 1.10, have been presented in recent years [LOH 12, GRO 08a, GRO 08b, AKA 09, ABE 06]. While the classic transistor is limited by a slope below the threshold of 60 mV/decade, suspended gate transistors have demonstrated subthreshold slopes of 2 mV/decade, thereby enabling static leakage to be limited [ABE 06].

Chapter 2 of this book will explore the most important transduction principles (actuation and detection) used on this scale.

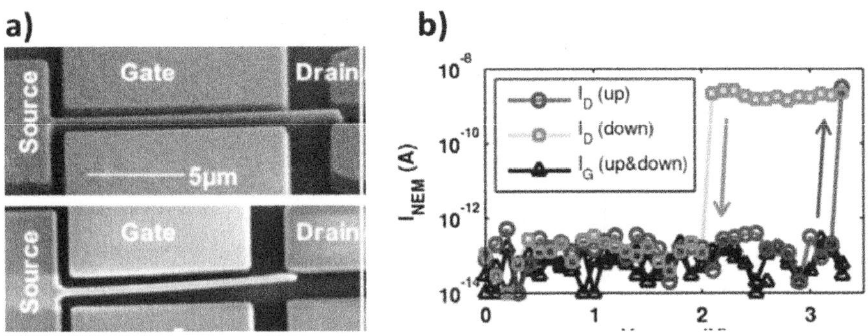

Figure 1.9. *Example of a nanorelay realized with a suspended nanocantilever [CHO 09]: a) the nanorelay seen through a scanning electron microscopy (SEM); b) example of current measurement: ON state when the beam is stuck to the drain electrode, OFF state without current when the beam is relaxing. Leakage is 5×10^{-14} A at the OFF state, which is several orders of magnitude lower than in current MOSFET transistors. For a color version of the figure, see www.iste.co.uk/duraffourg/nems.zip*

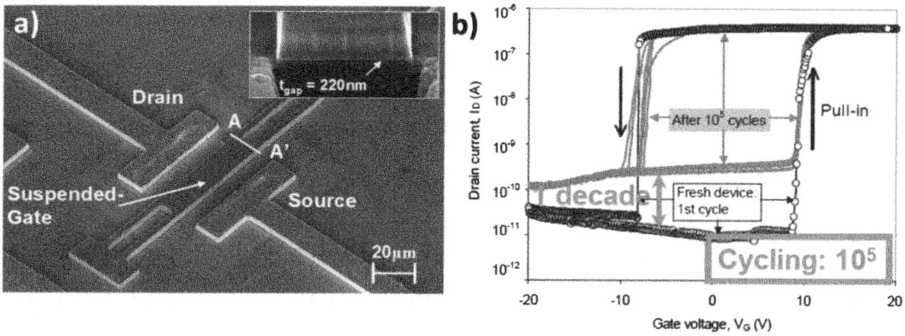

Figure 1.10. *Suspended gate MOSFET (SG-MOSFET): a) a beam acts as a suspended gate positioned above an MOS transistor channel; b) the current is modulated according to the position of this gate – in high position, the transistor is switched off and in low position (the gate is stuck to the channel) the transistor is on. Leakage is 10^{-11}A then 10^{-10} after 10^5 cycles*

NEMS are also very good tools for observing mesoscopic phenomena [BLE 04]. A large number of teams are specifically looking to measure the final displacement of a vibrating nanobeam at the quantum limit [SCH 05]. This limit corresponds to the fundamental vibrational quantum state of NEMS. A beam kept at high frequency oscillation ω_0 and at low temperature ($\hbar\omega > k_B T$) is a quantum oscillator and can thus be described by the formalism of quantum physics (Euler–Bernouilli's classic mechanical equation will be quantified). Heisenberg's uncertainty principle predicts that the position of a beam cannot be known with greater precision than $\Delta x_{SQL} = \sqrt{\hbar/2m\omega_0}$ where m is the actual mass of the beam and ω_0 is its resonance eigen-frequency [BRA 92]; Δx_{SQL} is called the standard quantum limit. In other words, the measurement disrupts the state of the system we are looking to measure. Two problems arise when we seek to reach this fundamental limit. It is necessary to cool the system such as to reach the quantum regime. Typically for a resonator with a frequency of 1 GHz, the temperature must be lower than 50 mK. This temperature cannot be reached using classic cryogenic means. An ultra-low noise detection scheme must be used and perturbation of the quantum state of the system must be kept to a minimum. The measurement system has a back-action effect on the quantum nanosystem and vice versa. This effect tends to modify the resonance frequency of the quantum nanoresonator, its local temperature and finally its amplitude via the modification of the damping. This permanent interaction between the detector and the quantum system (the cooled NEMS) sets the detection limit just slightly above the standard quantum limit [CAV 82]. It is also possible to use back-action to locally cool the NEMS below the cryogenic room temperature in order to reach the fundamental state.

The displacement of the quantum resonator can be measured by realizing a capacitive coupling between the nanoresonator and a mesoscopic system such as a single-electron transistor (SET) [KNO 03, LAH 04, NAI 06] (see Figure 1.11), a quantum box or quantum points of contact [CLE 02a]. The beam constitutes an electrostatic gate controlling the electrons leaping from one point to another. Another method consists of using superconducting quantum interference device (SQUID) micromagnetometers, one branch of which is constituted by the vibrating beam [ETA 08, BLE 08, POO 11].

Physicists have also been inspired by work on cooling atoms that exchange their energy through an anti-Stokes interaction with photons of laser sources [COH 98] to both reach the fundamental state of the resonator[1] and read its movement. The mechanical system is, therefore, coupled with an optical resonant cavity with an extremely high level of fineness. According to an early measurement diagram, the microsystem is replaced with one of the mirrors of a Fabry–Pérot cavity or is included in the optical cavity. The measurement is, therefore, interferometric and the optomechanical interaction is realized via the pressure force exerted by the photons over the mechanical element [SCH 06, KIP 08, GRÖ 09, MAH 12, GIG 06]. In the second approach, the mechanical system interacts via the evanescent coupling with integrated optical microresonators such as rings or photonic crystals [ANE 09, SCH 08, SCH 09, LIN 09, LIN 10]. Similarly, the nanoresonator can be coupled with a microwave resonant cavity [NAI 06, O'CO 10, HER 09]. These optomechanical systems will be presented in Chapter 4.

Mesoscopic physics has also been applied to the domain of quantum information using Cooper-pair boxes [NAK 99, VIO 02] or micro-SQUIDs [SCH 00a]. For the former, the qubit is constituted of the superposition of quantum states of excess charge in a superconducting island (coupling is electrostatic), whereas for the latter, the qubit arises from the superposition of states of flow (coupling is magnetic). Recently, the team of K. Schwab

1 Back-action cooling of the light on the resonator.

coupled a Cooper-pair box with a nanoresonator to read quantum states of charge [LAH 09]. For a complete description of superconducting qubits and their application in quantum calculus, please refer to [BLA 03]. In the same vein, nanomechanical structures have been put in oscillation via a smart transduction based on the charge shuttle [GOR 98, ERB 01]. Different experimental attempts to measure the quantum heat brought by a phonon have also been conducted [SCH 00b, FON 02].

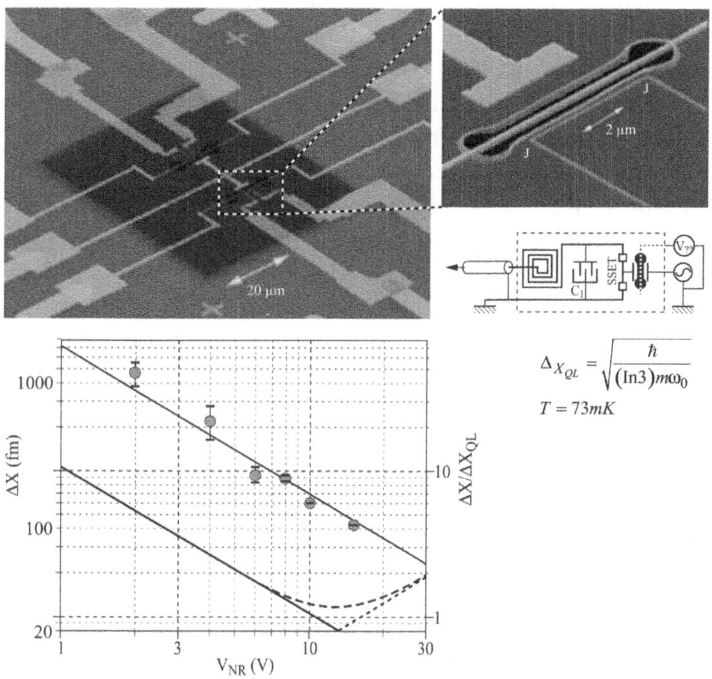

Figure 1.11. *Final measurement at low temperature of the displacement of an oscillating beam at 20 MHz according to Lah et al. [LAH 04] – a SET is used as an electrometer with a sensitivity of ~10 µe/√Hz – black curve: the sound of gunfire – red curve: the sound of the counter reaction – blue curve: the square sum considering uncorrelated sources of noise. For a color version of the figure, see www.iste.co.uk/duraffourg/nems.zip*

1.2. Conclusion

NEMS are interesting both in terms of fundamentals and applications. They often have high mechanical resonance frequencies (typically from 1 to 100 MHz) and dissipate low quantities of energy (mechanical and electrical).

They are sensitive enough to enable mass measurements to be realized at the single molecular level (molecule counting), to count electrons or phonons one by one, or to measure forces approaching the pico-Newton. This statement can be backed up by writing the expressions of the main mechanical properties of a nanobeam dependent on a reduction coefficient α applied similarly to its three dimensions. The expressions and the typical values of these mechanical parameters are summarized in Table 1.1. It can be seen that reducing the scale results in a linear increase in the resonance frequency when α decreases. The mechanical and thermal time constants decrease depending on the scale factor according to the linear and square laws. In other words, NEMS respond to requests as fast as the size of the nanobeam is small. A nanowire 1 μm long for a 50 nm^2 section in silicon can, therefore, thermalize in less than 10 ns. It is, therefore, possible to actuate high frequency NEMS with a thermomechanical force, something that was not possible with microsystems. This also means that a NEMS sensor can be used to detect fast phenomena (~1 μs or less)[2].

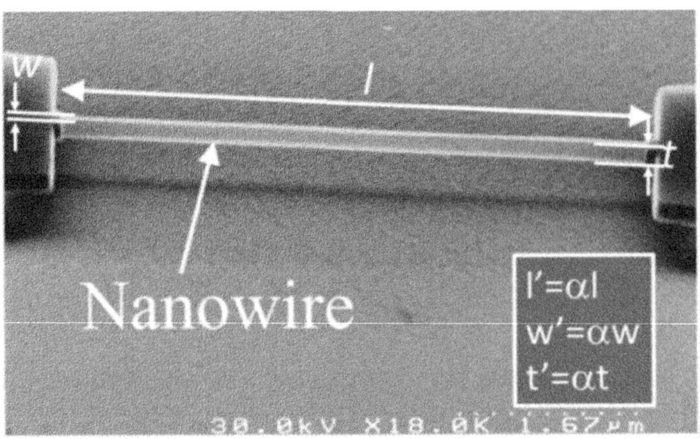

Figure 1.12. *Suspended silicon nanowire with a length l of (3 μm), a width w of (80 nm) and a thickness t of (160 nm) (source: LETI)*

2 The overall response time of the sensor must, however, take monitoring electronics (actuation and detection) into consideration. In some cases, this part will be slower than the NEMS itself.

Parameters	Law		Typical values
Mass	$m \propto wlt$	α^3	1 pg–10 fg
Stiffness	$k \propto Ewt^3/l^3$	α	1 N/m–10^2 N/m
Frequency	$f \propto \sqrt{E/\rho}\, t/l^2$	α^{-1}	10 MHz–1 GHz
Dissipated mechanical energy	$P_{th} \propto 2k_BTQ/\pi f$	α	100 aW–10 fW
Mechanical time constant	$\tau_m \propto Q/2\pi f$	α	0.1–10 µs
Thermal time constant	$\tau_{th} \propto c\rho l^2/\kappa$	α^2	0.1–100 ns
Noise amplitude	$\sqrt{2k_BTQ/fk}$	1	1–100 fm
Noise in force	$\sqrt{2k_BTk/fQ}$	α	10 fN–1 pN
Mass detection limit	$\delta m = 2m\, \delta f/f_0$	α^3	(ag – yg)

Table 1.1. *Orders of magnitude of the main electromechanical characteristics of a nanowire and the associated scaling laws when a reduction coefficient α is applied to its length, width and thickness: $l'=\alpha l$, $w'=\alpha w$, $t'=\alpha t$ (Figure 1.12) – E, ρ, c, κ are the Young's modulus, density, the thermal capacity and thermal conductivity of the nanowire, respectively. k_B and T are the Boltzmann constant and temperature, respectively*

2

Transduction on the Nanometric Scale and the Notion of Noise

How can an efficient electrical transduction of a mechanical displacement be realized on the nanometric scale? To respond to this question, we will consider a nanoelectromechanical system (NEMS) that is resonant to its eigenfrequency. A large number of nanosensors use frequency detection, as was described in the previous chapter (e.g. see Figure 1.8(a)).

Studying resonant sensors is made much easier if the NEMS can be described in terms of its transfer functions, each describing an elementary transformation (see Figure 2.1). In a linear regime, these conversions are represented by the transfer functions K_A and K_D. The movement of the resonator caused by actuation is represented by a mechanical transfer function denoted as $\alpha(\omega)$ and which was presented in Figure 1.8(d). A brief reminder of the mechanics will therefore be given in order to describe this mechanical response and the related notions.

Regardless of the application, it is imperative that the movement can be detected. Nevertheless, to design a well-performing nanosensor, it is necessary to evaluate the noise and the signal-to-noise ratio (SNR) of the overall system made up of the NEMS, the transduction means and the readout electronics. This chapter will therefore conclude with the study of the sources of noise for a number of previously mentioned detection schemes.

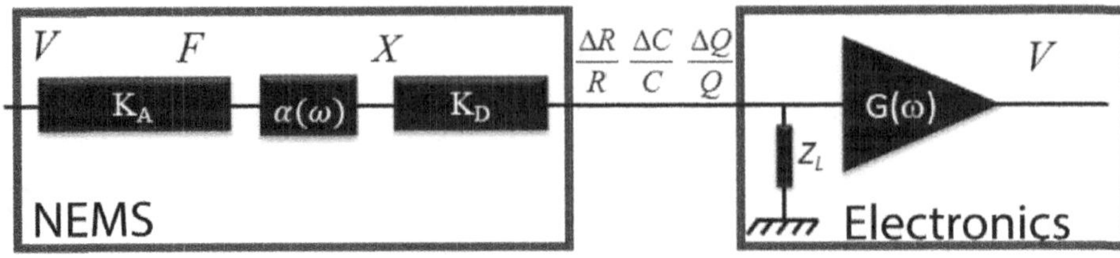

Figure 2.1. *Operating scheme of an electronic reader: the NEMS is represented by a transfer function product (linear hypothesis). A first transfer function K_A turns the conversion of an actuation voltage into a force (electrostatic, thermal, piezoelectric, etc.). The second $\alpha(\omega)$ makes the NEMS move as a result of this force. The third transfer function K_D turns the variation of an electrical characteristic (impedance, charge, etc.) into a response to the mechanical request. An electronic readout circuit converts this variation into output voltage. Z_L is the input impedance of the electronics seen by the NEMS*

2.1. Mechanical transfer function

For illustrative purposes, we are going to detail the method for determining the (Lorentzian) transfer function of a simple beam, which is the most commonly used geometry for NEMS. In particular, we are going to try to calculate the resonance frequencies f_n of this beam, as well as the associated effective masses[1]. For lengths over vibrating thickness ratios much larger than 1 and for small displacements, the movement of the beam can, at any point, be described by the Euler Bernouilli equation [WEA 90]:

$$EI\frac{\partial^4 y(x,t)}{\partial x^4}+b\frac{\partial y(x,t)}{\partial t}+\rho S\frac{\partial^2 y(x,t)}{\partial t^2}=F(x,t) \qquad [2.1]$$

where E is the Young's modulus, I is the square moment of inertia $I = wt^3/12$ (considering a movement according to the direction of the thickness), ρ is the density, $S = wt$ is the section of the beam, l is its length, b is the fluid damping coefficient, x is the position along the beam, where it is clamped at the origin, $F(t,x)$ is the actuation force by length unit and $y(t,x)$ is the deformation along the beam according to time.

Galerkin's decomposition method is used to resolve this equation. It consists of expressing $y(t,x)$ on the basis of the modal functions or the eigenmodes of the beam. The following variable separation is also used:

---

1 Mass in movement during vibration.

$$y(t,x) = \sum_n y_n(t)\varphi_n(x) \qquad [2.2]$$

where $\varphi_n(x)$ are the mechanical eigenmodes – shape of the beam – and $y_n(t)$ are the temporal vibrations. The functions $\varphi_n(x)$ are solutions to the equation at the eigenvalues below and form an orthonormal basis:

$$\frac{\partial^4 \varphi_n}{\partial^4 x} = \lambda_n^4 \varphi_n \qquad [2.3]$$

where λ_n are the eigenvalues associated with the modal functions $\varphi_n(x)$ – eigenvalues of the linear operator $\partial^4/\partial^4 x$. The eigenfunctions meet the following normalization condition:

$$\frac{1}{l}\int_0^l \varphi_n(x).\varphi_m(x)dx = \delta_{mn} \qquad [2.4]$$

where $\delta_{mn} = 1$ if $m=n$, or 0 otherwise.

The eigenvalues λ_n and the eigenmodes $\varphi_n(x)$ are defined by equation [2.3] by imposing the mechanical conditions on the limits of the following general solution:

$$\varphi_n(x) = A_n\, sinh(\lambda_n x) + B_n\, cosh(\lambda_n x) + C_n\, sin(\lambda_n x) + D_n\, cos(\lambda_n x) \qquad [2.5]$$

These conditions to the limits are defined by the way in which the beam is held at its extremities, at $x = 0$ and $x = l$ (e.g. clamped, pinched, positioned and free). Let us take the example of a cantilever, where the beam is anchored at $x = 0$ and free at $x = l$:

$$\begin{aligned} &y_n(x=0,t) = 0 \quad \frac{\partial^2 y_n}{\partial^2 x}(x=l,t) = 0 \\ &\frac{\partial y_n}{\partial x}(x=0,t) = 0 \quad \frac{\partial^3 y_n}{\partial^3 x}(x=l,t) = 0 \end{aligned} \qquad [2.6]$$

These equations are true whatever instant t is being considered and impose no constraint on the temporal part of the function $y_n(x,t)$. If the beam is double-anchored (or clamped), the conditions to the limits become:

$$\begin{aligned} &y_n(x=0,t) = 0 \quad y_n(x=l,t) = 0 \\ &\frac{\partial y_n}{\partial x}(x=0,t) = 0 \quad \frac{\partial y_n}{\partial x}(x=l,t) = 0 \end{aligned} \qquad [2.7]$$

By inserting the expression of $y_n(x,t)$ in the conditions [2.6], we obtain a system of linear equations:

$$A_n + C_n = 0$$

$$B_n + D_n = 0$$

$$A_n(sinh(k) - sin(k)) + B_n(cosh(k) - cos(k)) = 0$$

$$A_n(cosh(k) - cos(k)) + B_n(sinh(k) + sin(k)) = 0 \quad [2.8]$$

where $k = \lambda_n l$.

This system results in what is known as a digitally determined transcendental equation, which can find out the values taken by the eigenvalues λ_n:

$$(sinh(k) - sin(k)).(sinh(k) + sin(k)) - (cosh(k) - cos(k))^2 = 0 \quad [2.9]$$

By way of example, the eigenvalues found by this equation [2.9] are shown in the table below:

	$\lambda_1 l$	$\lambda_2 l$	$\lambda_3 l$	$\lambda_4 l$
Cantilever	1.875	4.694	7.854	10.995
Doubly-clamped beam	4.73	7.853	10.995	14.14

Table 2.1. *Eigenvalues for the first four modes of a cantilever and a double-anchored beam*

To reach the complete expression of the normalized modal functions, the coefficients A_n, B_n, C_n and D_n must be calculated using equation [2.8] and normalized with equation [2.4]. The forms of the first modes for a doubly clamped beam are given in Figure 2.2.

To determine the resonance eigenfrequencies, we solve equation [2.1] without second member $F(t,x)$=0 and without damping b. To do so, each member of this equation is projected over the mode n by multiplying them by $\varphi_n(x)$ and integrating them over x between 0 and L :

$$\mathrm{EI}.\, y_n(t).\, \lambda_n^4 + \rho S \frac{\partial^2 y_n(t)}{\partial^2 t} = 0 \quad [2.10]$$

The temporal part is a sinusoidal function of the form $y_n(t) = Ae^{-j\omega_n t}$ with ω_n as the angular frequency of the nth mode. We therefore rewrite equation [2.10] in the Fourier domain:

$$y_n(t)\frac{EI.\lambda_n^4}{\rho S} - {\omega_n}^2.y_n(t) = 0 \qquad [2.11]$$

The resonance frequency of the nth mode is therefore easy to express:

$$f_n = \frac{\omega_n}{2\pi} = \frac{\lambda_n^2}{2\pi}\sqrt{\frac{EI}{\rho S}} = \frac{\lambda_n^2.t}{4\pi}\sqrt{\frac{E}{3\rho}} \qquad [2.12]$$

By inserting the eigenvalues in the latter equation, it can be stated that the resonance frequencies are proportional to the vibrating width t and inversely proportional to the square of the length l (contained in parameter λ_n). These resonance eigenfrequencies will naturally be higher in a doubly clamped beam than in a cantilever. It is also interesting to calculate the ratios between two successive frequencies, which come back to the square of the ratio between two successive eigenvalues (see Table 2.2 below).

	f_2/f_1	f_3/f_2	f_4/f_3
Cantilever	6.27	2.8	1.96
Doubly clamped beam	2.76	1.96	1.65

Table 2.2. *Ratio between two successive resonance frequencies for a cantilever and a doubly clamped beam*

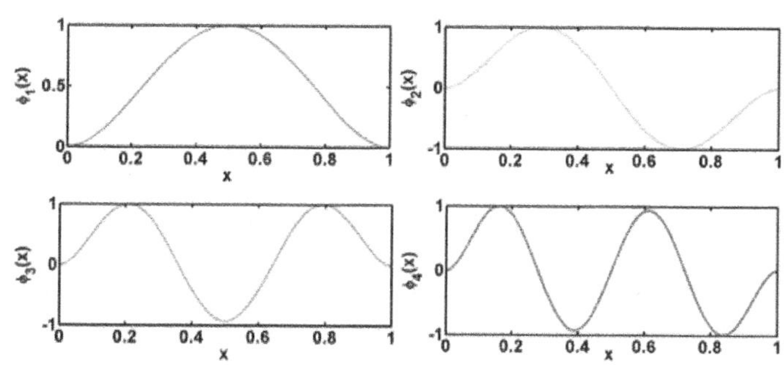

Figure 2.2. *Eigenmode along the beam with x normalized to its length for the first four resonance modes of a doubly clamped beam. The number of nodes and antinodes increases with the order of the modes*

Once the mechanical bases are in place, we will define the transfer function corresponding to the one-dimensional (1D) model equivalent to a beam. To do so, we need to once again pick up the Euler–Bernouilli equation with the second member, which is the actuation force. We will consider a sinusoidal force directed by a perpendicular direction to the beam and express it by separating the variables:

$$F(x,t) = g(x)f(t) = g(x).F_0.e^{-i\omega_d t}$$

where $F_0.g(x)$ is the force density per unit length – $g(x)$ varies between 0 and 1[2].

We will project equation [2.1] on the first mode to have a frequency response for an actuation on its first resonance mode at f_1:

$$EI.y_1(t).\lambda_1^4 L + \rho SL\frac{\partial^2 y_1(t)}{\partial^2 t} + bL\frac{\partial y_1(t)}{\partial t} = f(t)\int_0^l \varphi_1(x)g(x)dx \quad [2.13]$$

When $M=\rho SL$ and $\eta = \int_0^l \varphi_1(x)g(x)dx$, we get the well-known equation of a mass/spring/damping resonator:

$$\ddot{y}_1(t) + \dot{y}_1(t)\frac{b}{\rho S} + y_1(t)\frac{EI.\lambda_n^4}{\rho S} = \frac{f(t)\eta}{M}$$

$$\leftrightarrow \ddot{y}_1 + \frac{\omega_0}{Q}\dot{y}_1 + \omega_0^2 y_1 = \frac{F}{m_{eff}} \quad [2.14]$$

where y is the displacement of a point mass $m_{eff} = {}^{M}/_{\eta}$ attached to a spring with a stiffness $EI.\,\lambda_1^4$.

Through identification, we find ${}^{b}/_{\rho S} = {}^{\omega_0}/_{Q}$ and ${}^{EI\lambda_1^4}/_{\rho S} = \omega_0^2$. Q is the system quality factor which is the ratio between the stored mechanical energy and the energy lost in the cycle. Usually, NEMS have quality factors between 1,000 and 10,000 in a vacuum and 100 in the air.

Finally, regardless of the resonator considered (doubly clamped beam, cantilever and nanowire) with in-plane or out-of-plane movement, its

---

2 $g(x)$ is, for example, the image of lines of electrostatic fields (during a capacitive actuation) normalized to the maximum amplitude. This function takes edge effects into account, for instance.

dynamic can be described by the transfer function from the Fourier transform applied in [2.14]:

$$\alpha_{NEMS}(\omega) = \frac{y_1(\omega)}{F(\omega)} = \frac{1/m_{eff}}{-\omega^2 + \omega_0^2 + j\omega\omega_0/Q} \quad [2.15]$$

The transfer function $\alpha(\omega)$ [2.15] above resembles a Lorentzian whose fineness $\Delta\omega/\omega_0$ is inversely proportional to the quality factor $Q - \Delta\omega$. It is defined as the bandwidth for the maximum amplitude –3 dB. At resonance:

$$|\alpha(\omega_0)| = Q/m_{eff}\omega_0^2$$

$$(\omega_0) = \arg(\alpha(\omega_0)) = -\pi/2$$

$$\frac{\partial\phi}{\partial\omega|\omega_0} \sim -2Q/\omega_0 \quad [2.16]$$

It should be mentioned that the gradient of the phase around ω_0 is proportional to Q. This property will be important when we need to calculate the frequency noises of the oscillators realized from mechanical nanoresonators.

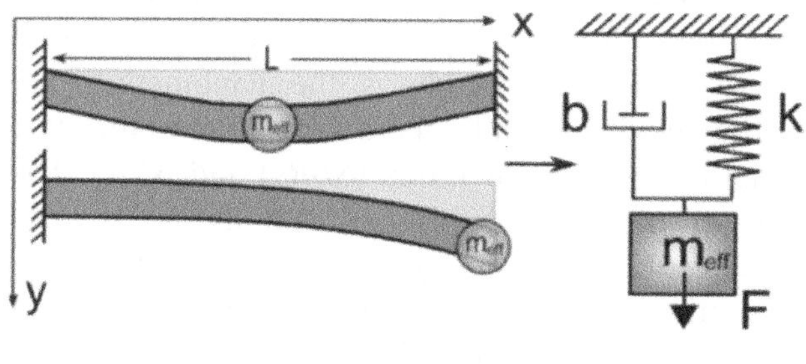

Figure 2.3. *1D model equivalent to a vibrating beam (cantilever or doubly clamped): the conditions at the limits are implicitly taken into account in coefficient η (or the stiffness k) and the expression of the force F which are functions of the eigenmode*

As the mechanical transfer function α, shown in Figure 2.1, has been defined, we must now find out the transfer functions corresponding to actuation K_A and detection K_D.

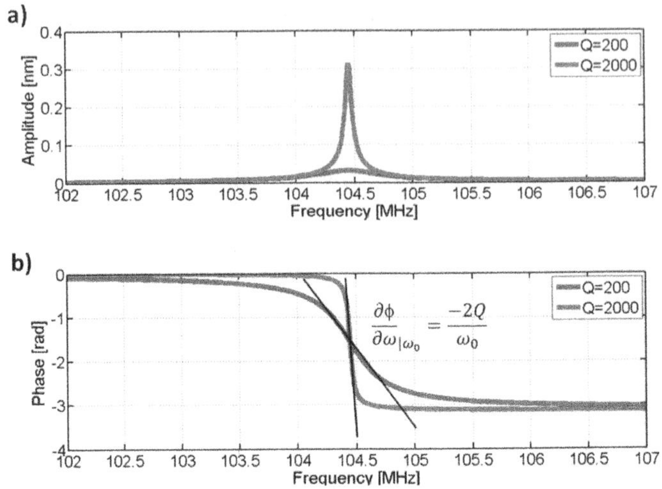

Figure 2.4. *Bode plot of the mechanical response of a resonator: a) amplitude for two quality factors Q (e.g. 2,000 in a vacuum and 200 in the air) and b) phase and expression of the gradient of the phase around the resonance frequency. For a color version of the figure, see www.iste.co.uk/duraffourg/nems.zip*

2.2. Transduction principles

This section will present the transduction principles that are commonly used in NEMS. The detection principles that accompany the reduction of size will be discussed in the next chapter, which is about the effects of miniaturization. How can we realize an efficient electrical transduction of a movement whose typical amplitude is the nanometer? The expression, "transduction efficiency" is usually heard in relation to the strongest possible continuous signal-to-background ratios (*SBRs*) and *SNRs*. *SNR* is the ratio between the power of the useful signal and the power of the noise. It characterizes the resolution of the system for a bandwidth of a given size. This important notion will be discussed in more detail at the end of the chapter. First, the useful electrical/electromagnetic signal must be higher than the continuous electrical background: in other words, we want *SBR*>>1. When measuring the vibrations of a nanoresonator, the continuous background term can seem improper as it is actually a radio frequency (RF) signal which comes, in part, from an electrical coupling between excitation and detection – the actuation signal being brought back over the output. Other sources, such as parasitic signals present in ground loops or radiating electromagnetic fields, can contribute to this continuous background (Figure 2.5).

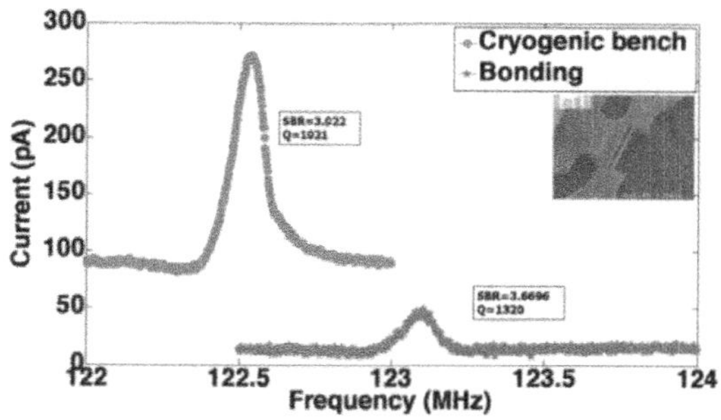

Figure 2.5. *Comparison of resonance signals (at 123 MHz) obtained with the same resonant silicon nanowire using piezoresistive detection (in insert) in two different measurement benches. In red: measurement bench in a vacuum and under tips, in green: measurement bench in a vacuum (a very similar vacuum) with connections via gold wires welded under the chip. The SBR moves from 2.5 (red) to 10 (green), showing the influence of the connection on the continuous background and the useful signal. For a color version of the figure, see www.iste.co.uk/duraffourg/nems.zip*

Moreover, when the output impedance of the NEMS is different from the input impedance of the measurement instrument – oscilloscope, network analyzer, lock-in amplifier (LIA) – the useful signal is dramatically reduced. It is therefore necessary to adapt the two impedances by adjusting them to the same value [EKI 02, TRU 07], which is generally close to 50 Ω with respect to RF standards. To illustrate the influence the input impedance of the readout circuit has on the signal, let us consider two microelectromechanical system (MEMS) [POU 03] and NEMS [DUR 08] sensors made up of a vibrating beam whose movement is detected through capacitance variation (see Figure 2.6). For the MEMS, the nominal capacitance is 53 fF for a variation of 2.5 fF. For the NEMS, these values are 115 and 5 aF, respectively. In comparison, the accumulated capacitance of the contact plots and the instrument input ranges from 1 to 100 pF, which is 100–10,000 times larger than the variation to be detected. By looking at the classic readout scheme shown in Figure 2.7, the output signal of these electronics will be divided by the simple effect of capacitive bridge, as shown in the equation below:

$$V_{out} \propto \frac{\delta C}{(C_{fb} + C_p + C_{gs} + C_0)} \qquad [2.17]$$

where δC is the variation of capacitance caused by the mobile beam, C_{fb} and C_{gs} are the capacitances of the gate and drain, respectively, and C_{gd} (with other capacitances potentially being added) is the gate/source at the first stage of the electronics. C_0 is the nominal capacitance of the NEMS, while C_p is the parasitic connection capacitance [COL 09a]. For a polarization voltage of 1 V, the output voltage will be proportional to 25 µV of MEMS and 50 nV in the case of NEMS. Classically, when using low-noise electronics, the continuous background ranges between 1 and 10 µV for a noise of 1 nV/√Hz brought back at input. In this context, it is therefore impossible to measure the vibration of the NEMS. The influence of the input impedances of the readout electronics and the parasitic capacitances is therefore exacerbated for NEMS. The main choice of which detection principle to use depends on the environment of the measurement.

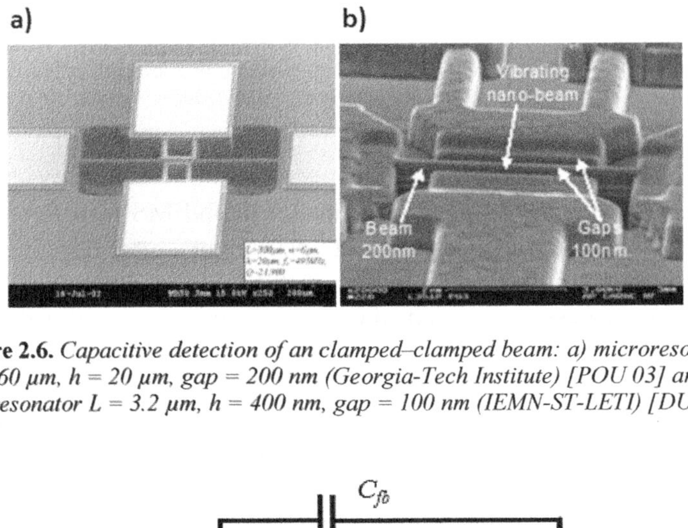

Figure 2.6. *Capacitive detection of an clamped–clamped beam: a) microresonator L = 60 µm, h = 20 µm, gap = 200 nm (Georgia-Tech Institute) [POU 03] and b) nanoresonator L = 3.2 µm, h = 400 nm, gap = 100 nm (IEMN-ST-LETI) [DUR 08]*

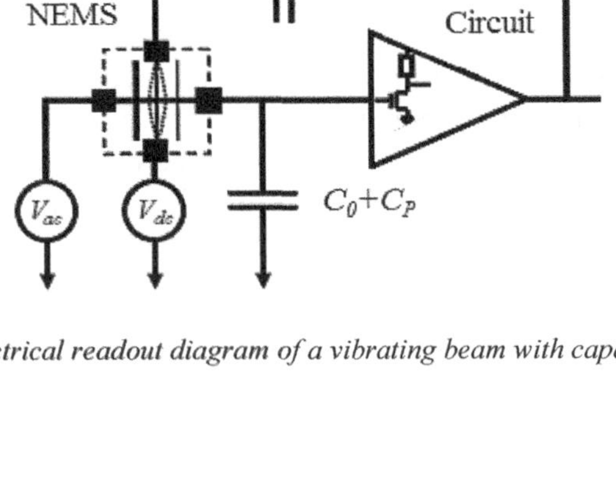

Figure 2.7. *Electrical readout diagram of a vibrating beam with capacitive detection*

Different measurement strategies have therefore been implemented to minimize this coupling and to limit the reduction of the signal [EKI 02, BAR 05, SAZ 04, GOU 10, ARC 10]. These instrumentation techniques will be discussed in more detail at the end of this section. Another approach also consists of bringing the electronics of nanosystems together in order to limit parasitic impedances [VER 08, VER 06a, ARC 12, OLL 12]. The most favorable way is to juxtapose the NEMS and its electronics complementary metal oxide semiconductor (CMOS in general) on one chip: this is known as monolithic integration and will be discussed in Chapter 3. Another smart alternative is to produce an active electromechanical component, such as the emblematic suspended gate MOSFET (SG-MOSFET), which was the first MEMS produced in the 1960s [NAT 67]. Naturally, its structure has been adapted for modern technologies [DUR 08a, KOU 13, GRO 08, ABE 06, BAR 12b, DUR 08b].

2.2.1. *The actuation of nanostructures*

Like for MEMS, common NEMS actuation principles are electrostatic, piezoelectric, magnetic or thermal. Mechanical nanostructures can also be put in movement by an optical force. To do so, the mechanical system is placed close to a waveguide in which light is guided. Even if the optical energy is confined to the inside of the guide, a small part of the wave will overflow this guide (evanescent wave). By placing the mechanical structure in its influence zone (several tens of nanometers to several hundreds of nanometers), the N/MEMS undergoes an optical force coming from the field gradient (E, H). In the same way as an electrostatic force, it derives from electromagnetic energy. The waveguide can be a simple guide (e.g. right), a ring-type optical resonant cavity [LI 08] or a photonic crystal [SAF 13]. This topic will be developed in Chapter 4, but for an overview of different methods and different optomechanical components, see [THO 10].

2.2.1.1. *Magnetic actuation*

Magnetic actuation relies on Laplace force which is exerted on a conductor passed by a current $i(\omega)$ and placed in a uniform magnetic field B (see Figure 2.8). This force is expressed according to the well-known law:

$$F(\omega) = i(\omega)LB \quad [2.18]$$

where L is the length of the current loop considered.

The expression of Laplace force can be defined from Lorentz force, which is the effort exerted on an electric charge in movement in an electromagnetic field (E, B).

Let us consider once again an NEMS with a doubly clamped nanobeam. The displacement at its center can easily be calculated using equation [2.15]:

$$y(\omega, {}^{L}/_{2}) = \phi_1({}^{L}/_{2}) \frac{i(\omega)B}{m_{eff}\left(-\omega^2+\omega_0^2+{}^{j\omega\omega_0}/_{Q}\right)} \quad [2.19]$$

Displacement depends linearly on the magnetic field B and the current i. When the field is uniform, $g(x) = 1$ in the expression of effective mass m_{eff} (see [2.13]) and the force per length unit is exactly F/L.

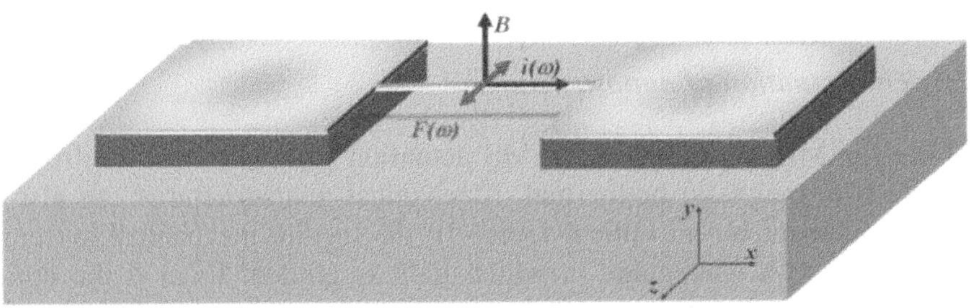

Figure 2.8. *Diagram showing Laplace force exerted on a conducting wire placed in a magnetic field and passed by an alternative current i(ω)*

This actuation is realized under a strong magnetic field, typically around 10 T. This requires the use of electromagnets in a cryogenic environment (77 K or less). This is an effective method that enables very small NEMS, such as nanowires [FEN 07, EKI 02, HUS 03], to be actuated at frequencies up to 1 GHz. To overcome implementation difficulties, it is possible to integrate a nanomagnet close to the NEMS. As the field of the magnet is much weaker, the distance between the magnet and the NEMS is 200 nm or less. The nanomagnet can be constituted of an alternation of ferromagnetic/antiferromagnetic layers, which are used to produce magnetic random access memories (MRAMs). The interaction between a ferromagnetic layer (the magnetic moments of domains are aligned and oriented in the same direction) and an antiferromagnetic layer (magnetic moments are aligned in antiparallel fashion) enables remanent magnetization to be obtained that is roughly equal to that at saturation, approximately 1–2 T. Depending on the stacks, the magnetic field can be in the plane of the current loop or perpendicular to it. To find out more details, consult [PAR 03]. These types of magnets can have lengths and widths less than 1 μm for the thickness of approximately 100 nm. An example of an NEMS actuated by a nanomagnet positioned below is given in Figure 2.9. The field B has three components B_x, B_y and B_z. In practice, only components B_x and B_z can actuate and generate a movement. As the out-of-plane movement in z is caused by the field B_x, as shown in the figure, this component is not uniform along the beam – it depends on the variable y – and the calculation presented in equation [2.19] becomes a bit more complex:

$$y\left(\omega, {}^{L}/_{2}\right) = \phi_1\left({}^{L}/_{2}\right) \frac{i(\omega) B_{0x} \int_0^L \phi_1(x) g(y) dy}{m\left(-\omega^2 + \omega_0^2 + {}^{j\omega\omega_0}/_{Q}\right)} \qquad [2.20]$$

The directions x, y and z are shown in Figure 2.8. Magnetic fields radiated at 100 nm above the magnet are 60 mT in the example chosen in Figure 2.10. When looking at a device such as the beam described in Figure 2.9, the resulting force F_z can nevertheless cause a vibration amplitude at the resonance $y(\omega_0, L/2)$ at the center in the order of several nanometers. To find out more details, refer to [BIL 09].

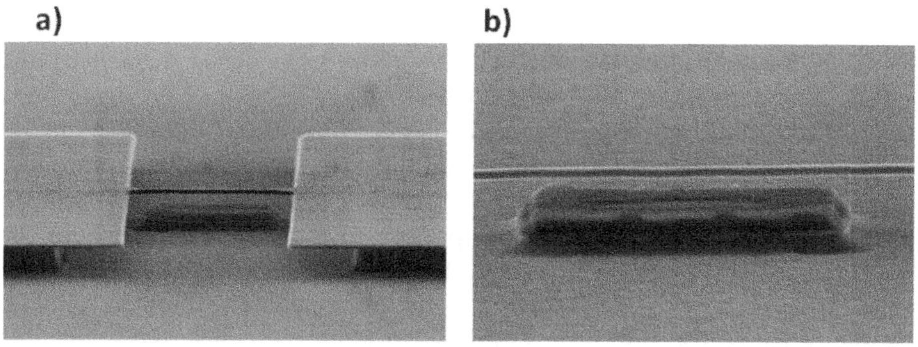

Figure 2.9. *Example of a nanobeam actuated by a Laplace force helped by a nanomagnet positioned below (gap of 100 nm): a) overview: platinum beam 2 μm long, 500 nm wide and 100 nm thick (image obtained from an electron microscope) and b) zoomed image of the magnet*

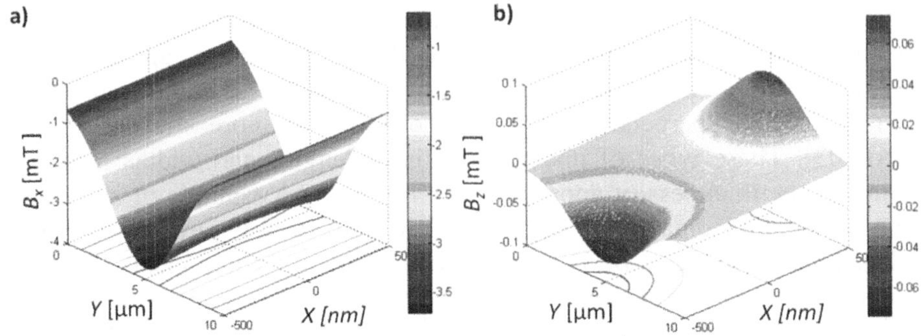

Figure 2.10. *Magnetic fields radiated by a nanomagnet in a beam positioned at 100 nm: a) component B_x and b) component B_z. For a color version of the figure, see www.iste.co.uk/duraffourg/nems.zip*

2.2.1.2. *Electrostatic actuation*

Electrostatic actuation is a widespread method of putting an MEMS or an NEMS in movement. This purely attractive force derives from the electrostatic energy contained between the two conducting surfaces concerned. By looking at the notations and the axes given in Figure 2.11, the density per unit length of the force is expressed as follows:

$$f(t,x) = \frac{C'(x,t)V_G^2(t)}{2} = C_n \frac{\varepsilon_0 t V_G^2(t)}{2(d-x)^2} \quad [2.21]$$

C' is the derivative in relation to x of the capacitance formed by the electrode and the beam, V_G is the actuation voltage applied to the electrode, d is the electrostatic gap and t is the thickness of the beam. C_n is a ponderation coefficient reflecting the edge effects on the field lines when d is no longer negligible than the thickness of the electrodes t. The edge effects tend to increase the real capacitance in comparison to a simple calculation of the capacitance plane (see [NAK 90] and [LEU 04]).

By considering the low displacements ($x<<d$), the force can be approximated at the order of 0 by $f(t,x) \sim C_n \frac{\epsilon_0 S V_G^2(t)}{2d^2}$ and displacement at the center of the beam is (according to [2.15]):

$$y(\omega, L/2) = \phi_1(L/2) \frac{C_n \frac{\epsilon_0 t V_G^2(\omega)}{2d^2}}{m_{eff}\left(-\omega^2 + \omega_0^2 + j\omega\omega_0/Q\right)} \quad [2.22]$$

The effective mass is calculated according to the width of the electrode l_a, that is, $\eta = \int_e^{e+la} \phi_n(x)dx$. C_n is considered as a constant value for small displacements. For example, let us consider nanowires (see Figure 2.12) that are 3 μm long, 80 nm wide (vibrating thickness for a vibration in the plane) and 160 nm thick. The electrostatic gap is 100 nm and the electrode has a length corresponding to 80% of the length of the wire. Force is estimated to several nano-Newtons, resulting in a displacement on the nanometric scale at the resonance ($Q \sim 1{,}000$).

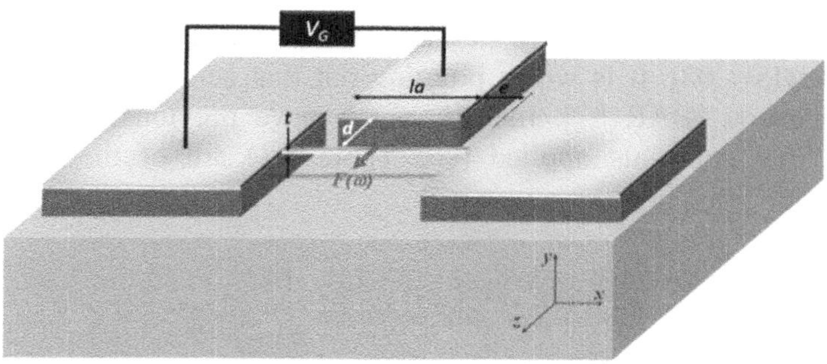

Figure 2.11. *Electrostatic actuation initiated by a voltage V_G applied between the beam and a (gate) electrode*

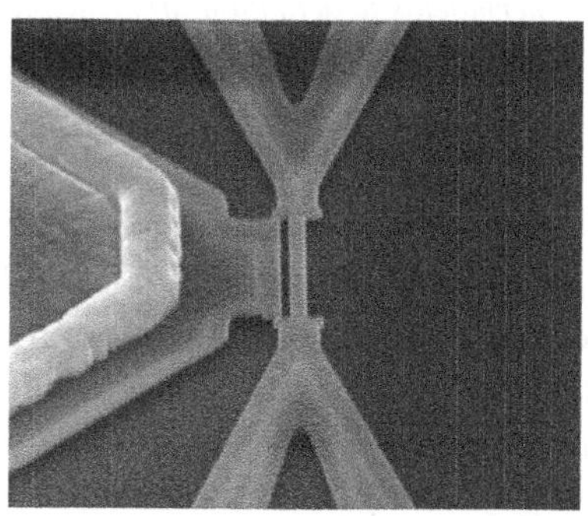

Figure 2.12. *Electrostatic actuation of a silicon nanowire: V_G is the actuation voltage applied to the gate and the wires are 160 nm thick by 80 nm wide and 3μm long*

Electrostatic actuation is simple, easy to integrate and varies in d^{-2}, which is favorable for reducing size. This scheme remains, however, sensitive to parasitic capacities, which greatly limits bandwidth well below the gigahertz. In contrast, this sort of limitation does not exist with magnetic actuation.

2.2.1.3. *Thermoelastic actuation*

Thermoelastic actuation is another option that has been widely used to put mechanical microstructures in movement. The actuation principle is based on the difference in thermal expansion coefficients of two materials stacked one on top of another, the set forming what is known as a bimorph. When the structure is heated, the difference in stress between the two layers causes the beam to deform (see Figure 2.13). Usually, the bimorph consists of a metallic layer on a beam, made up, for example, of silicon. Heating therefore occurs by polarizing the metallic loop via the Joule effect. Displacement $y(\omega)$ can easily be calculated using Timoshenko's calculus [TIM 25, HSU 02]. It is worth pointing out that actuation frequencies can reach gigahertz [BAR 07]. Indeed, as we have previously discussed, the thermal time constants applied (see Table 1.1) are of the order of the nanosecond.

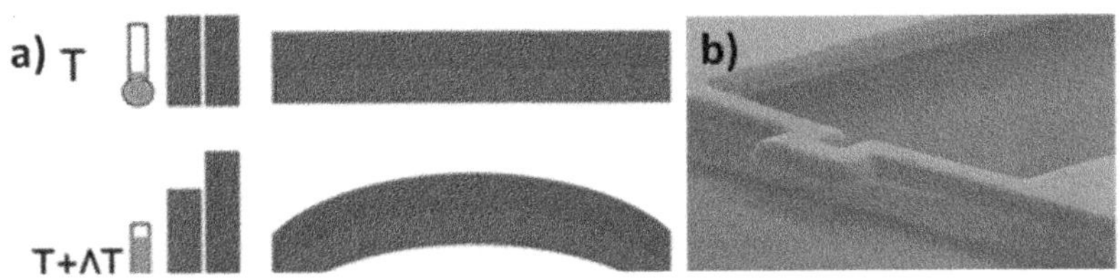

Figure 2.13. *Thermoelastic actuation: a) principle schema and b) example of a structure under thermoelastic actuation: a metallic loop (gold layer) on a silicon cantilever enables actuation*

One of the main problems of this type of actuation is electrical consumption, which can be high, in particular for NEMS array (e.g. 100 × 100 NEMS). Consequently, the rise in temperature can be a redhibitory factor in certain applications relating to the mass sensor [EKI 04]. In this event, a too-high temperature causes a strong desorption rate [YON 88] of the masses, which are physically absorbed at the surface of the NEMS, thereby preventing measurement.

2.2.1.4. *Piezoelectric actuation*

The final mode of actuating MEMS/NEMS is piezoelectric. Piezoelectric actuation uses specific dielectric layers which polarize when a mechanical constraint is applied (direct piezoelectric effect) or deform when an electric field is applied (inverse piezoelectric effect). The direct effect is used to actuate the mechanical element and the inverse effect to detect its movement. In a classic dielectric material, the polarization P of the material[3] varies linearly with the external electric field E: $P = \varepsilon_0[\chi]E$, where $[\chi]$ is the dielectric susceptibility. The polarization of the material originates from the creation of dipolar moments due to electron displacement under electrostatic force. On the one hand, there is the electronic dipolar moment which depends on the frequency of the applied electric field. Let us consider a molecule occupying a site in the crystal lattice of the material placed in an oscillating electric field. At very high frequencies (in the ultraviolet domain), the electronic cloud of the molecule deforms (electronic resonance similar to the resonance of a spring-mass system whose mass is charged). Infrared causes the molecule kernels to vibrate, whereas microwaves cause them to rotate. In the case of ionic crystals, in addition to electronic polarization, there is also ionic polarization born from the relative displacement of the charges in the crystal lattice. For a detailed description of the mechanisms, see [KIT 98]. In a piezoelectric material, polarization also depends on the mechanical constraint that is applied. This phenomenon is the result of the crystal symmetry of the dielectric and can only exist in 20 symmetry groups out of the 32 groups. The groups with a center of symmetry (centrosymmetrical) are not piezoelectric.

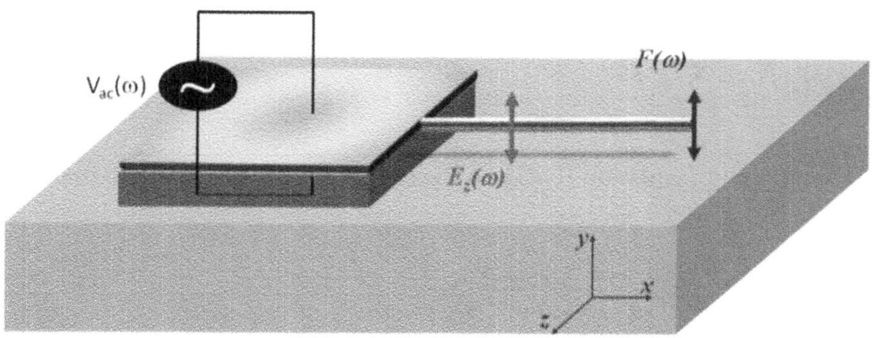

Figure 2.14. *Piezoelectric actuation – operating principle*

In MEMS technology, PZT, AsGa and AlN are commonly used materials. Their thicknesses can vary from 10 nm to several microns depending on the researched deflections and the disposition technologies. For NEMS, thin layers of 50 nm or less are favored [IVA 11] in a bid to

3 The polarization of a material is the dipolar moment of the material per unit of volume (average value of microscopic dipolar moment per unit of volume).

preserve the nanometric size of the system. In the example presented in Figure 2.14, a metal stack (Mo – 100 nm)/piezoelectric layer (AlN – 50 nm)/metal (Mo – 100 nm) is put on a mechanical part (in silicon or SiN insulating material). The electric field E_z in the thickness of the cantilever causes an elongation ε_{xx} of the longitudinal axis of the piezoelectric layer, such that $\varepsilon_{xx}=d_{31}E_z$, where d_{31} is the transverse piezoelectric coefficient (connecting the electric field to the constraint). The difference in elongation causes shear constraints leading the cantilever to be deformed from top to bottom depending on the sign of the electric field $E_z(\omega)=V_{ac}(\omega)/t_{AlN}$.

The calculation of deflection to resonance is very similar to that shown in equation [2.14]. Nevertheless, actuation is not expressed through a force but with a piezoelectric moment $M(x,t)$, causing bending [IVA 11]:

$$\langle EI\rangle_{eq}\frac{\partial^4 y(x,t)}{\partial x^4}+b\frac{\partial y(x,t)}{\partial t}+\rho w\frac{\partial^2 y(x,t)}{\partial t^2}=\frac{\partial^2 M(x,t)}{\partial x^2} \qquad [2.23]$$

where ρ is the mass density per surface unit (and not per volume unit), $\langle EI\rangle_{eq}$ is the equivalent rigidity of the multilayer (Metal/AlN/Metal/Si) and w is the width of the cantilever:

$$\langle EI\rangle_{eq}=\sum_i E_i I_i \qquad [2.24]$$

By projecting equation [2.23] on the reduced modal base of the first mode [IVA 11]:

$$y(\omega,{}^{L}/_{2})=\frac{wL\phi_1'(L)\beta_{piezo}V_{AC}}{m\left(-\omega^2+\omega_0^2+{}^{j\omega\omega_0}/_{Q}\right)} \qquad [2.25]$$

where β_{piezo} is a parameter that depends on the thickness of the layers, the position of the neutral fiber of the transverse piezoelectric coefficient e_{31} (linking the electric field to the bending), and ω_0 is the deformation eigenfrequency, which is defined in equation [2.14]. The product EI must, however, be replaced by its equivalent counterpart $\langle EI\rangle_{eq}$. m is the total mass of the cantilever.

Looking at equation [2.25], it can be seen that piezoelectric actuation varies linearly with actuation voltage. The efficiency of the actuation is chiefly set by the coefficient β_{piezo}. It should also be noted that this actuation requires the use of multilayers with internal constraints that must be controlled. These constraints can also limit the quality factor value Q.

We will now focus on the methods for detecting the displacements $y(\omega)$ discussed here according to the actuations selected. Detection methods play a particularly important role as measurement noise is caused by the detection element. Its control electronics, which depends on the transduction chosen, also has an impact on the level of output noise.

2.2.2. *Detection*

Some detection principles are similar to those used in MEMS. In particular:

– magnetic [FEN 07];

– capacitive [TRU 07];

– by field effect (SG-MOSFET) [NAT 67, DUR 08a, DUR 08c COL 09b];

– piezoresistive [MIL 10, HE 08, KAN 82];

– piezoelectric [KAR 12];

– detections.

Other detection principles more specific to NEMS have been created in the past decade:

– detection by field emission [AYA 07, JEN 07];

– detection by junctionless transistor (JLT) [KOU 13, BAR 14];

– detection with a resonant channel (VB-FET) [BAR 12b, BAR 12c, GRO 08];

– piezoresistive detection using giant piezoresistive effects [HE 08, HE 06, KOU 11];

– piezoelectric detection by 2D electron gas [TAN 02];

– optical detection by evanescent coupling with an optical microresonator[4] [ANE 09, ASP 13].

To evaluate the effectiveness of a detection method, the simplest method consists of writing its related scaling law (see Figure 1.12). We will therefore follow this pragmatic, though reductionist, approach. Indeed, no detection technology can be dissociated from the readout circuit of the output signal (by reflectometry, in transmission with the array analyzer, by heterodyning, etc.). In other terms, the essential characteristics presented at the beginning of this chapter, *SNR* and *SBR*, can only be estimated when the instrumental method is taken into consideration. We will therefore complete the scaling law obtained by several examples of experiments from the literature in order to think about the analysis obtained from the simple readout of this scaling law. These laws also offer the opportunity to discuss the most commonly used instrumental methods.

2.2.2.1. *Magnetic detection*

The first detection principle for a movement with very low amplitude was magnetic detection. It is based on Lenz's law: when a beam passed by a current is in movement in a magnetic field (which occurs when the beam is deflected by Laplace force, as described in equation [2.20]), electromotive voltage V_{EMF} appears at its terminals:

$$V_{EMF} = -\frac{\partial \varphi(t)}{\partial t} = -\frac{\partial}{\partial t}\iint \vec{B}.\overrightarrow{dS} \qquad [2.26]$$

where $\varphi(t)$ is the magnetic flow cut by the movement of the beam at an instant t.

This voltage is largely dependent on the geometry of the beam used. Picking up [2.19] for a uniform magnetic field [FEN 07]:

$$V_{EMF}(t) = -B\int_0^L \phi_1(x)dx\frac{dy_1}{dt} \qquad [2.27]$$

4 Optomechanical detection using the coupling of a mechanical system and an optical resonant cavity was initially developed to enable gravitational waves to be detected in the big interferometers VIRGO (Europe) and LIGO (US). Refer to [PIN 00] and [COU 01]. This is the "inverse" effect of that used to optically actuate microstructures.

Let:

$$V_{EMF}(\omega) = \phi_1\left({}^{L}/_{2}\right)\frac{j\omega\int_0^L \phi_1(x)dxLI(\omega)B^2}{m_{eff}\left(-\omega^2+\omega_0^2+{}^{j\omega\omega_0}/_{Q}\right)} \qquad [2.28]$$

Rewriting the equation gives:

$$V_{EMF}(\omega) = \frac{j\omega\gamma L^2 I(\omega)B^2}{M_{eff}\left(-\omega^2+\omega_0^2+{}^{j\omega\omega_0}/_{Q}\right)} \qquad [2.29]$$

with $\gamma = \int_0^L \phi_1(x)dx$ and $M_{eff} = m_{eff}/\phi_1\left({}^{L}/_{2}\right)$. When considering a doubly clamped beam, $\gamma = 0.81$ and $M_{eff} = 0.76m$. The output voltage linked to the vibration of the nanobeam is proportional to the squared magnetic field.

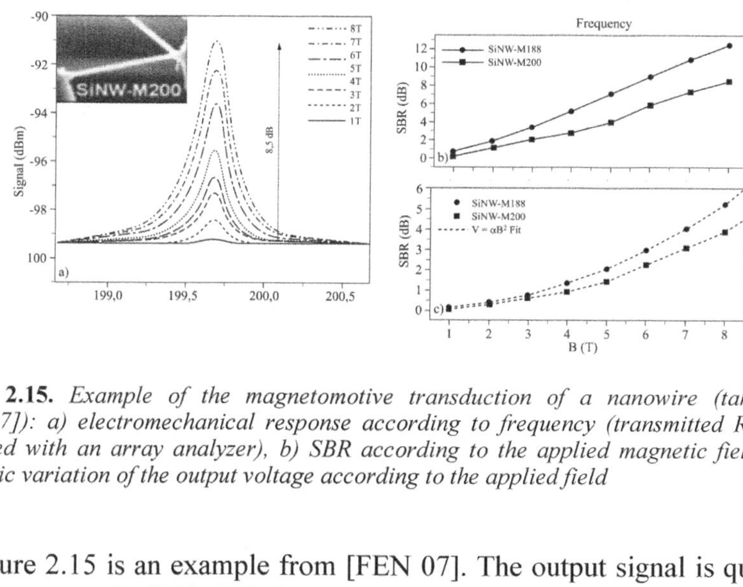

Figure 2.15. *Example of the magnetomotive transduction of a nanowire (taken from [FEN 07]): a) electromechanical response according to frequency (transmitted RF power measured with an array analyzer), b) SBR according to the applied magnetic field and c) quadratic variation of the output voltage according to the applied field*

Figure 2.15 is an example from [FEN 07]. The output signal is quadratic with the magnetic field. SBR is 10 dB under a strong field of approximately 8 T. Considering equation [2.28] and supposing a current I independent of the size of the beam, magnetomotive detection at the resonance follows the following scaling law:

$$[V_{EMF}(\omega)] = \left[\frac{Q\gamma L^2 I(\omega_0)B^2}{M_{eff}\omega_0}\right] \propto \frac{\alpha^2}{\alpha^3\alpha^{-1}} = 1 \qquad [2.30]$$

The reduction in size does not, in principle, aid detection. Nevertheless, it is an efficient and simple detection that was used in first intention with strong magnetic fields exceeding 8 T [FEN 07].

2.2.2.2. *Capacitive detection*

Capacitive detection is another detection used in first intention. The signal can be rapidly reduced by the presence of parasitic capacitance (see equation [2.17]). Let us consider an electrostatic actuation at the resonance described in Figure 2.11. By applying a readout voltage V_l at the terminals of the variable capacitance, the output voltage is expressed as:

$$V_{capa}(\omega) = \frac{V_l \delta C(\omega)}{C_0 + C_p} = \frac{V_l C_0}{C_0 + C_p}\frac{y(\omega)}{d} = \frac{V_l C_0}{C_0 + C_p}\phi_1\left(L/2\right)\frac{Q C_n \epsilon_0 t V_G^2}{2 d^3 m_{eff} \omega_0^2} \quad [2.31]$$

It can be seen that the output voltage is proportional to the readout voltage and is quadratic with the actuation voltage V_G. The scaling law can be obtained from [2.31]:

$$\left[V_{capa}(\omega)\right] \propto \alpha^{-2} \quad [2.32]$$

The reduction in size appears to greatly benefit the transduction scheme mixing electrostatic actuation and capacitive detection. In practice, the capacitive bridge will destroy this benefit, unless to limit the parasitic capacitance C_p as much as possible (plot capacitance, interconnection, etc.).

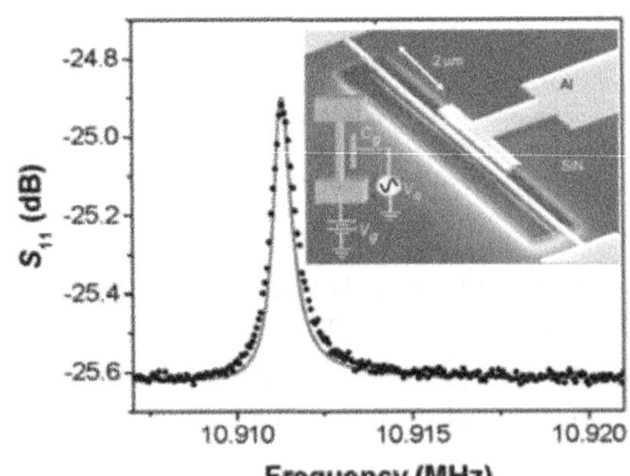

Figure 2.16. *Example of electrostatic actuation and capacitive detection of an SiN nanobeam (from [TRU 07]) – electromechanical response according to frequency (reflected power measured by an array analyzer)*

A resonator with capacitive detection can be modeled on an equivalent electric circuit $R_mL_mC_m$ [NGU 99] corresponding to what are known as motional impedances, i.e. impedances linked to current variation i_m caused by mechanical vibration $y(\omega)$. Figure 2.16 is an example of electrostatic actuation and capacitive detection realized on a vibrating nanobeam at 11 MHz. The instrumental method is based on a direct transmission readout scheme of the signal detected by an array analyzer. To limit the effects of the capacitance bridge, an impedance adaptation circuit at resonance frequency is placed between the component and the array analyzer. Despite this, *SBR* is lower than 1 dB, thus showing the limits of this detection above 10 MHz. As will be discussed later in the book (see Chapter 3), this requires the use of electronics placed as close as possible to the nanoresonator.

2.2.2.3. *Detection by transistor effect: SG-MOSFET*

The final monolithic integration is the SG-MOSFET component. It is composed of a bending vibrating beam in the plane (horizontally) and doped zones forming the source, channel and drain of an MOS transistor in the thickness of the plate. The beam is used as the transistor gate. The gate dielectric is composed of an air gap whose value varies with the displacement of the beam and an oxide, as shown in Figure 2.17. The bending of the gate causes a variation in the electrostatic surface potential of the channel. The quantity of inversed charges is therefore modulated along the channel by the mechanical oscillations of the SG. The resulting drain current varies proportionally to the amplitude of its movement. This type of detection was, in fact, invented by Nathanson [NAT 67] and has come back into fashion in recent years [DUR 08b].

The component was developed using two distinct technological approaches: silicon on nothing (SON), with a silicon thickness of 400 nm, or silicon on insulator (SOI), with a silicon thickness of 160 nm (Figure 2.18). The technology for SON was joint-created by LETI and the company STMicroelectronics. In both SON and SOI, a 5 nm-thick oxide is deposited on the side of the channel. The SG is actuated by an independent MOSFET electrode. Electrostatic actuation is set by the difference in potential (V_g-V_{dc}) and by sinusoidal voltage alternating current (AC) V_{ac}. The operating point of the transistor is regulated by voltages V_g and V_d in order to be in saturation regime to maximize the output current and have the greatest sensitivity possible at the movement of the beam. To aid comprehension, the voltages applied are described in the 3D diagram in Figure 2.17.

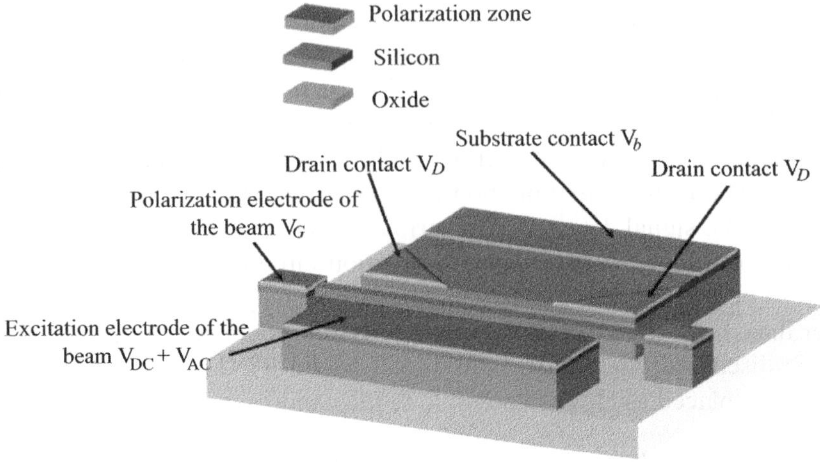

Figure 2.17. *Electrostatic actuation and detection by SG-MOSFET – operating scheme. For a color version of the figure, see www.iste.co.uk/duraffourg/nems.zip*

If you are feeling brave, the physical model of the component is described in [DUR 08a]. The theoretical results of this model have been compared to measurements from the SG-MOSFET device [DUR 08b]: the experimental[5] and theoretical parameters S_{12} are presented in Figure 2.19 for purely capacitive devices (see Figure 2.6(b)) and for SG-MOSFET devices in Figure 2.18(a). The parameter S_{12} is measured by an array analyzer. The continuous SBR (from the analyzer) is 8 dB for SG-MOSFETs compared to 2 dB for classic capacitive detection, showing that active SG-MOSFET detection presents a slight improvement. In principle, these detection modes are very similar. Indeed, the advantage presented by the MOSFET, or its transconductance g_m, is controlled by the surface potential that varies according to the vibration of the beam. Given that this potential depends on the variation in capacitance between the vibrating beam and the channel, the scaling laws remain unchanged. In practice, it is required to find a compromise between the operating point, which sets the vibration of the mechanical part, and the operating point, which sets transconductance. Nevertheless, it is difficult to find the optimal voltages in order to have a strong mechanical response and the optimal voltages in order to have a strong electric response, as they are often distant.

5 S_{12} is the output power to RF input power ratio expressed in decibels.

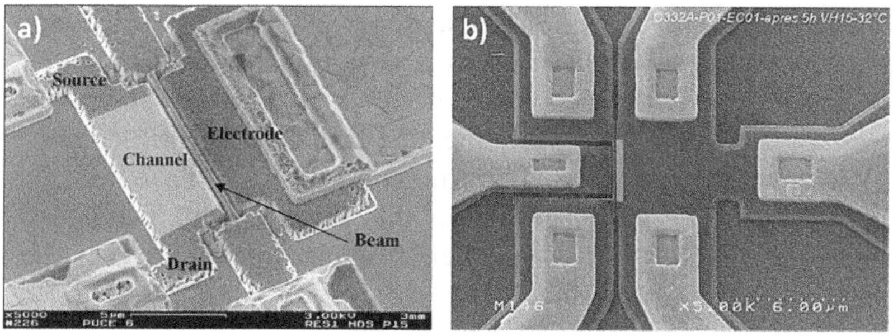

Figure 2.18. *SG-MOSFET – the gate is a suspended beam that is free to move: a) SON technology (thickness = 400 nm, gate length = 5 μm, gap between the gate and the channel = 120 nm, width of the channel = thickness) [COL 09] and b) SOI technology (thickness = 160 nm, gate length = 1 μm, gap between the gate and the channel = 100 nm, width of the channel = thickness) (source: LETI). For a color version of the figure, see www.iste.co.uk/duraffourg/nems.zip*

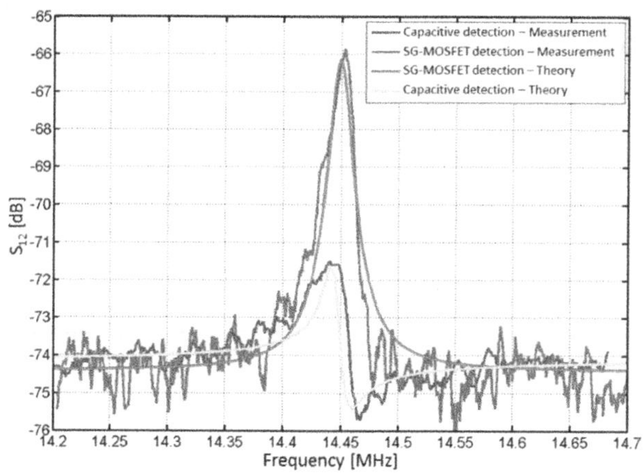

Figure 2.19. *Parameter S_{12} – theoretical results comparing SG-MOSFET and capacitive NEMS of the same sizes versus measurements from experiments. For a color version of the figure, see www.iste.co.uk/duraffourg/nems.zip*

2.2.2.4. *Piezoresistive detection*

One of the important detection principles is based on the piezoresistive effect. The electrical resistance of a piezoresistive material varies according to a mechanical constrains[6] (see equation [2.33]). This characteristic has

6 Traditionally, this constraint is axial.

been widely used for pressure sensors in MEMS technology. The piezoresistive part, which is used as a constraint gauge, is located at the anchoring of the membrane, which deforms as a result of pressure. The piezoresistive element can be a metallic layer or surface-doped silicon. NEMS with what is known as "piezometallic" detection have been described in the literature [LI 07]. They have the advantage of being low noise (Johnson noise is very low), but their response remains weak: the gauge factor [2.33] has just a few units.

When considering the notations given in Figure 2.20, the relative resistance variation $\Delta R/R$ is expressed as follows:

$$\frac{\Delta R}{R} = \pi_L \sigma_L + \frac{\Delta L}{L}(1+2\nu) = \left[(E\pi_L) + (1+2\nu)\right]\frac{\Delta L}{L} = G\varepsilon_L \qquad [2.33]$$

where σ_L is the applied axial constraint, π_L is the axial piezoresistive coefficient linking the resistance variation to constraint (Pa^{-1}), ν is the Poisson coefficient and G is the gauge factor linking the relative resistance variation to the elongation (without unit). Its expression contains a purely geometric term indicating a section change through the Poisson coefficient and a characteristic term of the material which is the coefficient π_L. G reduces in geometric terms for metals, explaining their low gauge factor.

In semiconductors, material term π_L is dominant. This can be explained intuitively as follows: any constraint applied to the silicon breaks the crystal cubic symmetry, which changes the structure of the energy bands. For doped silicon N, this changes the amount of electrons in conduction bands. The number of electrons with an effective longitudinal mass can, for example, reduce in favor of the number of electrons with an effective transverse mass (and inversely according to the direction of the constraint). The conductivity of the silicon therefore changes according to the constraint applied. In the case of doping P, the constraint tends to deform the valence bands and raise in particular the degeneration at the center of the first Brillouin zone at $k = 0$. The effective masses of the holes find themselves modified. In other words, the piezoresistivity of silicon P originates from the deformation of valence bands, whereas in silicon N, it stems from the modification of the distribution of the carriers (see Figure 2.21). This is an anisotropic phenomenon and the values of the piezoresistive coefficients π_L depend on the crystal axis being considered. For more information, see [BAR 09], [RIC 08] and [HER 54]. In a macroscopic silicon crystal for crystal orientations of <100>, the coefficient G is 100. In the case of nanowires constructed with the bottom-up technique and a doping P, a giant effect has been observed and G can reach 5,000. This scaling effect has exclusively been observed in the case of doping P. We will return to this phenomenon in Chapter 4.

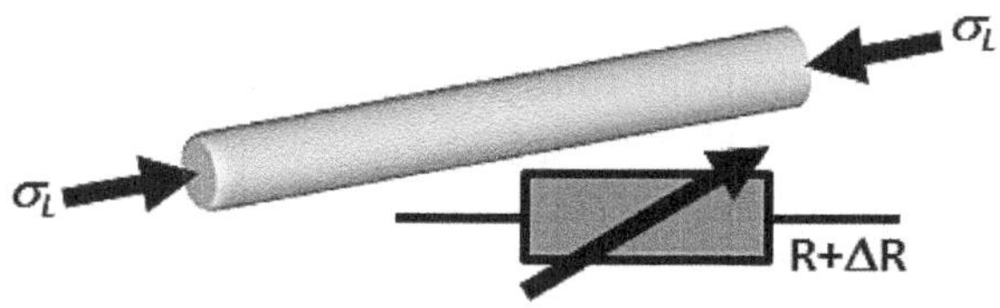

Figure 2.20. *Diagram of a wire constrained by an axial force (mechanical stress and compression)*

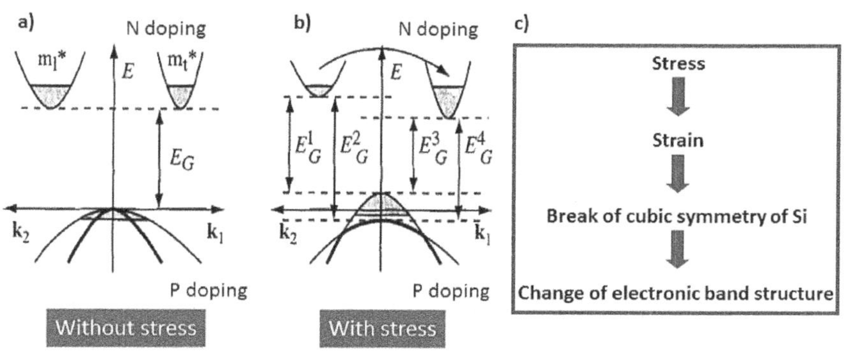

Figure 2.21. *Intuitive explanation of the piezoresistive effect in silicon: a) drawing of the structure of the silicon band without constraint, b) deformation of bands under constraint and c) intuitive explanation of the piezoresistive effect in silicon*

Let us look at Figure 2.11, which describes a doubly clamped nanobeam (or nanowires) actuated by an electrostatic force. Let us imagine that the movement caused is detected using the piezoresistive effect. When displacement is large enough, the bending of the beam generates an axial constraint that causes the variation of its electric resistance. We will now return to the Euler–Bernouilli equation [2.1], which determines the bending movement of a beam, and consider, in addition, the nonlinear constraint term:

$$EI\frac{\partial^4 y(x,t)}{\partial x^4} - \left(\sigma S + \frac{ES}{2L}\int_0^L \left(\frac{\partial y(\xi)}{\partial \xi}\right)^2 d\xi\right)\frac{\partial^2 y(x,t)}{\partial x^2} + b\frac{\partial y(x,t)}{\partial t} + \rho S\frac{\partial^2 y(x,t)}{\partial t^2} = F(x,t) \qquad [2.34]$$

With the conditions with the following limits: $\begin{cases} y(0)=0 \\ y'(0)=0 \\ y(L)=0 \\ y'(L)=0 \end{cases}$

σ is a potential axial constraint (initial constraint in silicon, by example) and the second term in brackets corresponds to the axial constraint of interest caused by the movement. The other terms have already been defined in equation [2.1]. The constraint $\frac{E}{2L}\int_0^L \left(\frac{\partial y(\xi)}{\partial \xi}\right)^2 d\xi$ described in equation [2.34] is the origin of the resistance variation ΔR due to the piezoresistive effect. By picking up equation [2.33], it becomes:

$$\frac{\Delta R}{R} = G\frac{\Delta L}{L} = \frac{G}{2L}\int_0^L \left(\frac{\partial y(\xi)}{\partial \xi}\right)^2 d\xi \qquad [2.35]$$

By considering the approximate form of the first eigenmode [POS 05] $\phi_1(x) = \sqrt{\frac{2}{3}}\left(1 - \cos(\frac{2\pi x}{L})\right)$, we can express the constraint term $\int_0^L \left(\frac{\partial y(\xi)}{\partial \xi}\right)^2 d\xi$ using the projection of y on this eigenmode [2.2] and thereby find an analytic expression to the relative resistance variation (ΔR/R):

$$\frac{\Delta R}{R} = G\frac{2\pi^2}{3\phi_1^2(\frac{L}{2})}\left(\frac{y}{L}\right)^2 = G\frac{\pi^2}{4}\left(\frac{y}{L}\right)^2 \qquad [2.36]$$

It is worth recalling that y is the vibration amplitude at the center such that: $y(\omega) = y(\frac{L}{2}, \omega) = y_1(\omega)\phi(\frac{L}{2})$. Resistance variation varies according to the square of the displacement of the beam; it is therefore a secondary effect.

For illustration purposes, we will refer to [KOU 13] to describe the piezoresistive detection of high-frequency (HF) vibrations of a silicon nanowire actuated by an electrostatic force. The wire is 30 nm in diameter for a length of 1.5 μm. The distance between the actuation electrodes and the wire is 80 nm (see Figure 2.22).

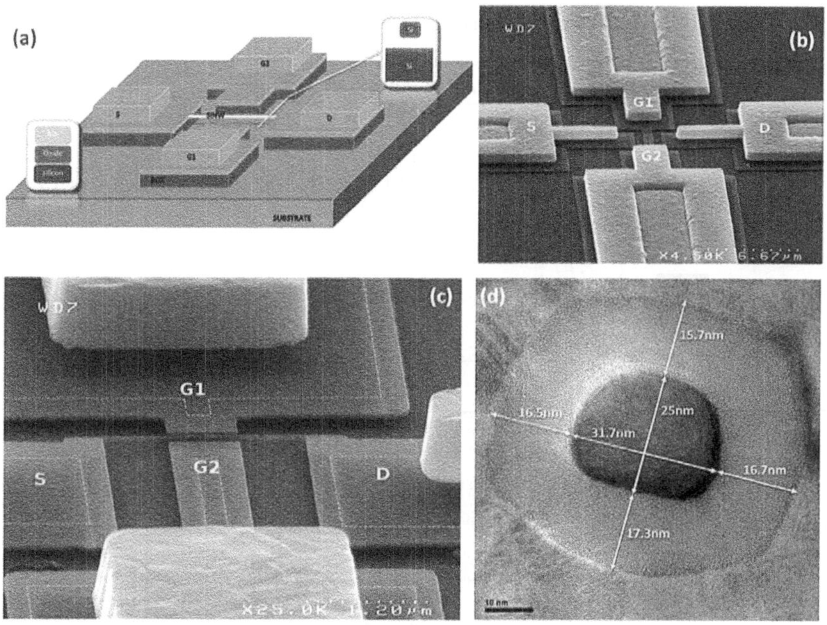

Figure 2.22. *Silicon nanowires 1.5 μm long and 30 nm in diameter: a) 3D schemes describing the stacking of the layers (gray: silicon, red: oxide and yellow: metal), b) view of the whole system seen through an scanning electron microscope (SEM), c) zoomed image of the nanowire and d) cross-section of a nanowire seen through a transmission electron microscope (TEM). For a color version of the figure, see www.iste.co.uk/duraffourg/nems.zip*

This example also enables us to illustrate a measurement scheme which is known as "down-mixing". This classic heterodyning method enables us to both read a low-frequency representative of the HF signal from the vibration and eliminate the continuous background (see Figure 2.23). In the previous section, we learned that displacement was proportional to the square of actuation voltage V_G [2.31]. In the case we considered, the nanowire is polarized by a voltage V_{DS} and actuated by a voltage V_G, which is the sum of a continuous component V_{DC} and a sinusoidal component $V_{AC}(\omega)$. The "electromechanical" current I_{DS} is therefore written as:

$$I_{DS}(\omega) = \frac{V_{DS}}{R(\omega)} = G\frac{V_{DS}}{R_0}\frac{\pi^2}{4L^2}\varphi_1^2(L/2)\left[\frac{C_n\frac{\epsilon_0 t V_{DC} V_{AC}(\omega)}{d^2}}{m_{eff}\left(-\omega^2+\omega_0^2+j\omega\omega_0/Q\right)}\right]^2 \quad [2.37]$$

Let $I_{DS} \propto V_{DS}(\omega - \Delta\omega) \times V_{AC}^2(\omega)$. R_0 is the nominal resistance of the wire. The current varies quadratically with V_{AC}, and the actuation frequency is set at $\omega_0/2$. At mechanical resonance, the peak current I_{DS0} is reduced to:

$$I_{DS0}(\omega_0) = G\frac{\pi^2}{4L^2}\frac{Q^2C_n^2\varepsilon_0^2t^2\varphi_1^2(L/2)}{m_{eff}^2\omega_0^4d^4}\frac{V_{DS0}V_{DC}^2V_{AC0}^2}{R_0} \quad [2.38]$$

V_{AC0} is the peak actuation voltage and V_{DS0} is the peak readout voltage.

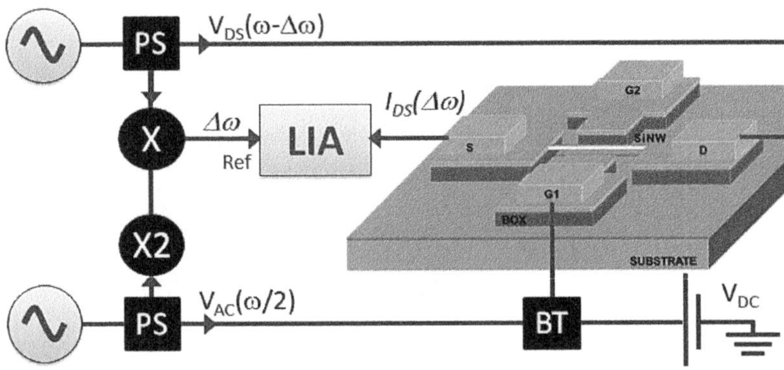

Figure 2.23. *Readout scheme by what is known as "down-mixing" heterodyning of a high-frequency vibrating nanowire: V_{DS} is the readout voltage whose frequency varies at $\omega \pm \Delta\omega$ in an interval centered on ω_0; V_{AC} is the actuation signal whose frequency varies around $\omega/2$. The nanowire behaves likes an RF mixer [2.37], the output current I_{DS} varies at the frequency $\Delta\omega$ and will be detected by a lock-in amplifier (LIA). BT stands for "Bias tee": this element enables an AC voltage to be added to a DC voltage*

The "down-mixing" heterodyning technique consists of modulating the readout signal V_{DS} at a frequency of $\omega \pm \Delta\omega$ and the excitation signal at a frequency of $\omega/2$ centered on $\omega_0/2$ (Figure 2.23). Equation [2.37] shows that the output current will therefore have a component $\Delta\omega$ accompanied by higher harmonics. This output current is detected by an LIA which enables a synchronous detection in order to keep only the low-frequency component $\Delta\omega$[7]. To reconstruct the electromechanical response HF, the excitation frequency is scanned over an interval centered on $\omega_0/2$. The readout signal is synchronized to excitation, in order to have a readout frequency that is constantly offset from $\Delta\omega$ during scanning. The peak value of the resonance current is given in equation [2.38]. A set of electromechanical responses dependent on frequency ω and peak excitation voltage is shown in Figure 2.24(a). The quality factor is 700. Figure 2.24(b) shows a quadratic behavior dependent on V_{AC}. The theoretical current is calculated with equation [2.38] by considering a gauge factor G of 236. The nominal resistance of the wire shown in Figure 2.22 is 1.2 MΩ. In this example, for an interplay of voltages V_{AC} = 100 mV, V_{DS} = 70 mV and V_{DC} = 300 mV, the axial constraint is ~120 kPa for a displacement y of approximately 1 nm. As shown in Figure 2.24, this leads to a signal to background ratio of 5 dB, which is quite low as the effect we are using here is second order.

According to equation [2.38], the scaling law is:

$$I_{DS0}(\omega_0) \propto \alpha^{-5} \qquad [2.39]$$

The scaling effect seems to be extremely beneficial for this transduction. Nevertheless, the nanowire rapidly reaches a nonlinear mechanical Duffing regime [NAY 04], contained in the constraint term $\frac{E}{2L}\int_0^L \left(\frac{\partial y(\xi)}{\partial \xi}\right)^2 d\xi$ of equation [2.1]. The critical amplitude A_c, above which the signal becomes nonlinear, can be roughly estimated with the formula $A_c = 1{,}685\frac{t}{\sqrt{Q}}$ [KAC 09]. By considering the quality factor Q to be 700 for a vibrating thickness t of 25 nm, the limit above which behavior is nonlinear is 1 nm. Above this amplitude, mechanical stiffness is no longer constant; rather it depends on the displacement. The narrower the wire, the faster this effect will appear.

7 Δω ranges from 10kHz to 100kHz (compared to resonance frequencies much higher than 10MHz).

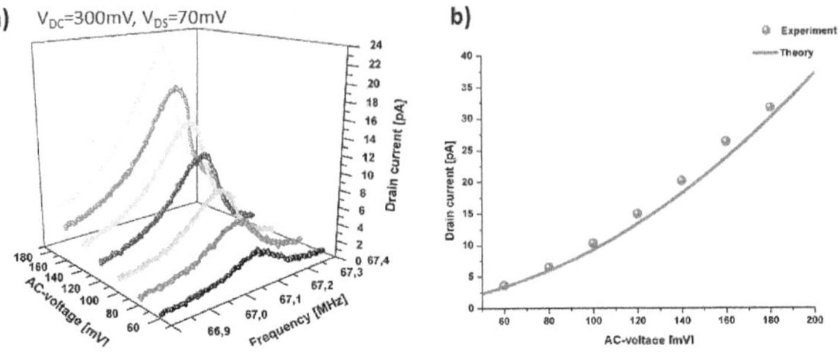

Figure 2.24. *Electromechanical response of a silicon nanowire obtained by piezoresistive detection in a "down-mixing" detection scheme. a) Response according to frequency and actuation voltage V_{AC}. b) Quadratic approximation of the output resonance frequency according to voltage V_{AC}. For a color version of the figure, see www.iste.co.uk/duraffourg/nems.zip*

We will now discuss in some detail Duffing mechanical nonlinearity. There are countless other nonlinear behaviors originating from mechanics or electromechanics [LIF 08]. Let us pick up the Euler–Bernouilli equation [2.34] and project this equation on the base $\varphi_1(x)$ using the same method described at the beginning of this chapter. We get an equation of a similar size to equation [2.14]:

$$\ddot{y}_1 + \omega_1^2 y_1 + k_d y_1^3 + \frac{\omega_1}{Q}\dot{y}_1 = f_1 \qquad [2.40]$$

where ω_1 (rad.s^{-1}) is the angular eigenfrequency and k_d is the Duffing coefficient (N.m^{-3}). Calculation results in the following expressions:

$$\omega_1 = \lambda_1^2 \sqrt{\frac{EI}{\rho S}\left(1 + \frac{L^2\sigma}{4\pi^2 EI}\right)}$$
$$k_d = \frac{ES}{2L}\left(\int_0^L \left(\frac{\partial \varphi_1(x)}{\partial x}\right)^2 dx\right)^2 = \frac{E(2\pi)^4}{18\rho L^4} \qquad [2.41]$$

Equation [2.40] can be resolved numerically. The amplitude of the mechanical vibration according to excitation frequency is shown in Figure 2.25 for a 2.5 μm nanowire and a square 40 × 40 nm^2 cross-section. For large displacements (higher than critical amplitude), the frequency response deforms at HFs, resulting in an increase in the mechanical stiffness of the nanowire. This effect can cause bistable, and later unstable, behaviors which, generally, are not desirable[8]. In conclusion, piezoresistive transduction calls on large displacements to cause a non-null second-order signal and its efficiency will in essence be limited by nonlinear mechanical effect that imposes a limit on the vibration amplitude around the critical amplitude.

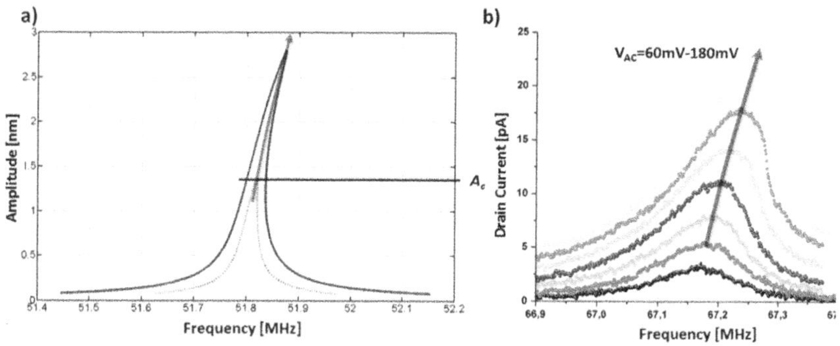

Figure 2.25. *Deflection of a silicon nanowire put in the resonance by an electrostatic actuation according to frequency. a) Mechanical vibration amplitude: (dotted curve) linear dynamic below the critical amplitude and (full curve) nonlinear dynamic with a hardening Duffing effect (order 3) deforming the peak of resonance toward high frequencies. b) Example of an experiment: electromechanical current detected on a nanowire made to vibrate by an electrostatic force. For a color version of the figure, see www.iste.co.uk/duraffourg/nems.zip*

To resolve this problem, an order 1 piezoresistive effect must be used. In other words, the constraint in the section of the nanowire must be directly generated by a compression or a voltage and not for a bending movement in large displacement. To this end, CEA-LETI suggested a design [MIL 10] called "X-beam" or "crossbeam", based on the lever arm effect. A color image of the X-beam is shown in Figure 2.26. Two nanowires placed on either side of a beam, not far from anchoring, are used as piezoresistive constraint gauges. The operating principle is as follows: the displacement of the nanocantilever generates a traction/compression force of nanowires which is proportional to the actuation force through a mechanical amplification coefficient (lever arm effect). The constraint caused in the

8 Some groups have worked on making two nonlinear effects compete (one hardening and the other softening) to increase the value of critical amplitude [KAC 09].

section of the wires will therefore cause a variation in the resistance by piezoresistive effect.

Figure 2.26. *SEM color image showing a crossbeam. The cantilever (in yellow) is 300 nm wide and 5 µm long. The lateral piezoresistive gauges (red on the left, green on the right), which collect the constraints, are 80 nm wide and 500 nm long. The system is 160 nm thick. F_{el} is the electrostatic actuation force. F_G is the force generated in the gauges (silicon is doped P at 10^{19} cm^{-3}). For a color version of the figure, see www.iste.co.uk/duraffourg/nems.zip*

The force applied to the nanogauges is expressed as follows [MIL 10]:

$$F_G(\omega) = \Gamma \frac{\omega_0^2}{\omega_0^2 - \omega^2 + j\omega\omega_0 / Q} F_{el}(\omega) \qquad [2.42]$$

where Γ is the amplification factor due to the lever arm. If the beam was totally rigid with a rotation point (instead of anchoring), this factor would be the ratio of the distance between the rotation point and the center of gravity of the beam and the distance between this same rotation point and the gauges. Constraint in the gauges is F_G/s, where s is the section of the gauges. Resistance variation ΔR can be expressed using the actuation force:

$$\frac{\Delta R(\omega)}{R} = G \frac{F_G(\omega)}{2 \cdot s \cdot E} \propto y_1(\omega) \qquad [2.43]$$

Resistance variation is directly proportional to displacement (see equation [2.15]). This is therefore a first-order effect, unlike the previous case which used a second-order effect. A greater efficiency of transduction can therefore be expected. Furthermore, in essence, the cantilever has a linear behavior over a larger displacement range, enabling stronger constraints in gauges to be reached in comparison to doubly clamped nanowires. To obtain the frequency response of the NEMS, a direct measurement (homodyne) can be made by electrostatically actuating the beam around its resonance and by polarizing the nanogauges by two different direct current (DC) readout signals. The output potential at the beam anchoring is measured. As the resistances of the gauges vary in the opposite direction, this differential measurement enables the background (the common mode) to be eliminated and the useful signal to be preserved. As for the simple nanowire, the use of a down-mixing measurement scheme (see Figure 2.23) is preferred in order to improve the SBR, especially when resonance frequencies are high. Several variants can be used:

1) The beam is actuated by a voltage V_G which is the sum of a continuous component V_{DC} and a sinusoidal component $V_{AC}(\omega)$. The frequency on either side of the resonance frequency ω_0 of the latter component is scanned. The nanogauges are polarized by two AC signals in phase opposition (to preserve the differential character of the system) at the excitation frequency with an offset $\Delta\omega$. The output voltage $V_s(\Delta\omega)$ is measured at the end of the beam by an LIA (see Figure 2.27(a)).

2) The beam is actuated by an AC voltage with a frequency of $\omega \pm \Delta\omega$ and the voltages from the nanogauges are measured by an LIA through a differential amplifier. In this instance, a low-noise amplifier from the LIA is used to our advantage to realize the differential and reduce the noise level (see Figure 2.27(b)).

3) and 4) The beam is actuated by an AC signal of $\omega/2$. As the electrostatic force follows a quadratic law with voltage (see [2.22]), the force will vary at a frequency double that of the actuation frequency ω (see Figures 2.27(c) and (d)). Frequently, a differential actuation is realized, whose aim is to eliminate the common mode (which results in background), caused by the capacitance present between the actuation electrodes and the vibrating beam (see Figure 2.27). For more information, refer to [ARC 10]. The output voltage is expressed according to the relative resistance variation:

$$V_s(\Delta\omega) \propto \frac{V_{Bias}(\omega)\Delta R(\omega-\Delta\omega)}{R} \qquad [2.44]$$

For an actuation with a continuous component, $V_{DC}+V_{AC0}cos(\omega t)$:

$$V_s(\Delta\omega) \propto \frac{QG\Gamma V_{Bias0} V_{DC} V_{AC0}}{2 \cdot s \cdot E} \quad [2.45]$$

For an actuation with a continuous component $V_{AC0}cos(\omega t/2)$:

$$V_s(\Delta\omega) \propto \frac{QG\Gamma V_{Bias0} V_{AC0}^2}{2 \cdot s \cdot E} \quad [2.46]$$

where V_{Bias0} is the peak readout voltage. The output voltage is proportional to the bias voltage and varies linearly or quadratically depending on the type of actuation being considered.

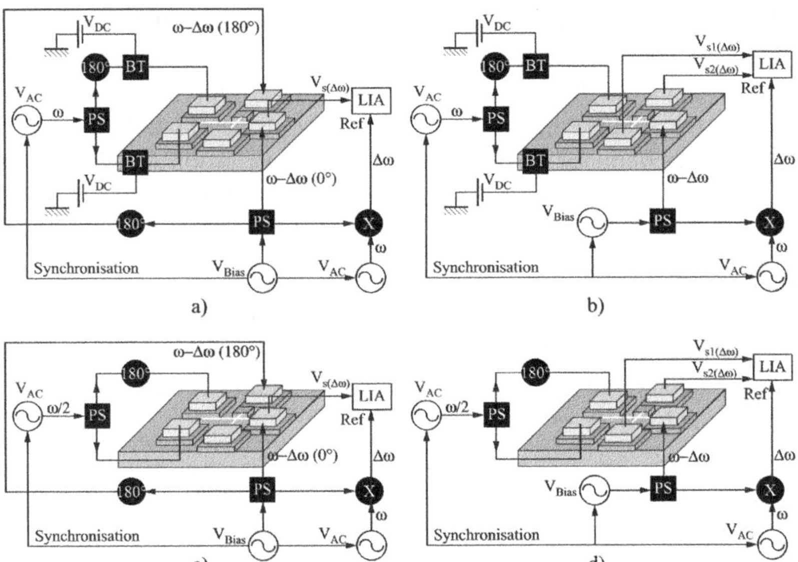

Figure 2.27. *Readout diagrams by heterodyning of an X-beam vibrating at high frequency ω (LIA: lock-in amplifier, PS: power splitter, symbol "X" mixer, symbol "180°": 180° phase shift; V_{DS} is the readout voltage whose frequency varies at $\omega \pm \Delta\omega$; the output voltage V_S varies at a frequency of $\Delta\omega$ and will be detected by the LIA. a) V_{AC} actuation signal whose frequency is set at ω The measurement is differential by taking the output signal over the anchoring of the beam. b) V_{AC} actuation signal whose frequency is set at ω The measurement is differential by using the differential amplifier of the LIA. c) V_{AC} is the actuation signal whose frequency is set at $\omega/2$. The measurement is differential by taking the output signal over the anchoring of the beam. d) V_{AC} is the actuation signal whose frequency is set at $\omega/2$. The measurement is differential by taking the output signal over the anchoring of the beam. It should be noted that in all instances, actuation occurs with two polarized electrodes in phase opposition. In so doing, actuation is differential, enabling the coupling capacitance between the electrode and the nanobeam to be eliminated, which results in undesirable background*

Figure 2.28 shows a typical resonance for a very small NEMS (beam: L=1.2 μm, w = 100 nm, t = 25 nm and gauges: l = 100 nm, s = 25 × 25 nm^2) measured with the readout diagram 4 (see Figure 2.27(d)). The nominal resistance R is 1.2 MΩ. For V_{Bias0} = 1.1 V, V_{AC0} = 0.8 V and V_{DC} = 0.3V, the displacement amplitude is estimated at 10 nm corresponding to an average constraint in the section s of a nanogauge of 1 MPa (i.e. one order of magnitude larger than for a doubly clamped nanowire). In these conditions, the SBR is 39 dB. This excellent *SBR* is also explained by actuation and differential detection, enabling the background to be drastically reduced. This example shows that beyond transduction efficiency, the measurement scheme must be carefully chosen.

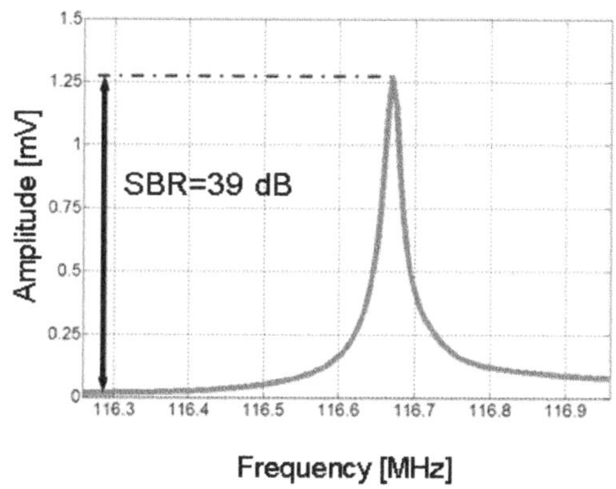

Figure 2.28. *Typical electromechanical response: output voltage V_s according to the actuation frequency ω scanned on either side of the resonance frequency ω_0*

Before concluding this chapter, we need to identify the detection noises and their impact on the resolution of frequency sensors when the latter are used in a loop system.

2.3. Self-oscillation and noises

We are going to study noise by considering an oscillating system that includes a mechanical nanoresonator (e.g. the nanocantilever shown in Figure 1.8), or a doubly clamped nanowire (see Figure 1.12). To implement this type of loop, the NEMS can be inserted either into a self-oscillating loop or a phase-locked loop. This architecture is less intuitive and will not be detailed here. As for all families of oscillators, their topology obeys the system description given in Figure 2.29. The electric signal from detection is amplified and phase shifted (pi/2) in order to meet oscillation conditions and is finally returned to an NEMS resonator via the actuation force.

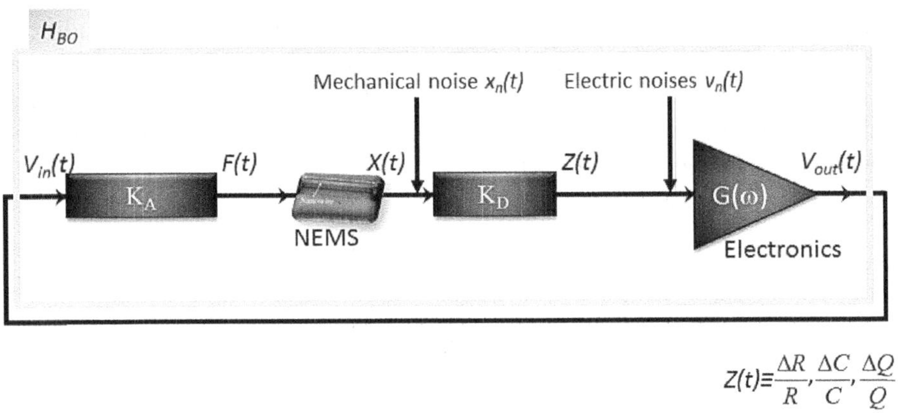

Figure 2.29. *Operating scheme of a closed-loop electric reader. The NEMS is shown with the transfer functions described in Figure 2.1 (linear hypothesis). The difference here lies in the loopback of the output of the electronics on the excitation of the NEMS, thereby forming a self-oscillating loop, if the Barkhausen conditions are encountered (see below)*

The oscillation conditions known as Barkhausen conditions that the electronics must meet are as follows:

$$H_{BO}(j\omega_0) = \frac{V_{out}}{V_{in}}(j\omega_0) = K_A(j\omega_0)\alpha(j\omega_0)K_D(j\omega_0)G(j\omega_0) = 1$$
$$\Leftrightarrow \begin{cases} \left\| H_{BO}(j\omega_0) \right\| = 1 \\ \arg\left(H_{BO}(j\omega_0) \right) = 0 \end{cases} \quad [2.47]$$

where $H_{BO}(j\omega_0)$ is the transfer function of the resonator associated with its transduction means in open loop (OL) and $G(j\omega_0)$ is the transfer function of the electronics to be sized. Furthermore, by considering transduction means as gains, we get phase and gain conditions that the electronics must meet to obtain self-sustaining oscillations at ω_0:

$$\begin{cases} \left\| H_{BO}(j\omega_0) \right\| = 1 \\ \arg\left(H_{BO}(j\omega_0)\right) = 0 \end{cases} \Leftrightarrow \begin{cases} \left\| G(j\omega_0) \right\| = \dfrac{m_{eff}\,\omega_0^2}{K_D K_A Q} \\ \arg\left(G(j\omega_0)\right) = -90° \end{cases} \qquad [2.48]$$

The reaction electronics must therefore phase shift the electric signal from detection by –90° while guaranteeing a sufficient gain in order to obtain self-sustaining oscillations (by counterbalancing the losses).

An oscillator is characterized by its phase noise, which causes a part of the fluctuation in frequency standard deviation σ_ω around its central frequency ω_0. By considering a control of the root mean square (RMS) amplitude $<x_c>$ of the oscillations, the phase noise of the oscillator can be expressed following the approach described in [EKI 04b] and [ROB 82]. σ_ω is directly linked to the noise spectral density in frequency S_ω by the Parseval–Plancherel theorem:

$$\sigma_\omega^2 = \int_{\omega_0-\Delta\omega/2}^{\omega_0+\Delta\omega/2} S_\omega(\omega)d\omega \approx S_\omega(\omega=\omega_0)\Delta\omega \qquad [2.49]$$

where $\Delta\omega$ is the measurement bandwidth required for self-oscillation by the bandwidth of the NEMS ω_0/Q. The overall performance of the NEMS and its control electronics depend on the integration time over which the measurement is made. In general, the noise spectral density is given for 1 Hz bandwidth corresponding to several signal periods. The frequency noise spectral density, which enables the spectral purity of the oscillator to be estimated, is expressed according to the phase noise spectral density – the conversion from phase into frequency is given by the gradient $\frac{\partial\phi}{\partial\omega}$ (Figure 2.4). The phase noise spectral density is proportional to the inverse of the SNR in OL [ROB 82, RAZ 96]:

$$S_\omega(\omega) = \frac{S_\varphi(\omega)}{(\partial\varphi/\partial\omega)^2} \approx \left(\frac{\omega_0}{2Q}\right)^2 \frac{S^x_{openloop}(\omega)}{A_c^2} = \frac{\omega_0^2}{4Q^2}\frac{1}{SNR} \qquad [2.50]$$

where $S_\phi(\omega)$ is the spectral density of the phase fluctuations of the oscillator (in dBc/Hz), $S^x_{openloop}(\omega)$ is the amplitude noise in OL and SNR is expressed in relation to power. It should be noted that equation [2.50] for phase noise is only valid in a very narrow band around resonance ω_0. To evaluate the phase noise over a wider band, Leeson's semi-empirical model should be prioritized [SAU 77].

$S^x_{openloop}(\omega)$ is the sum of the spectral densities of all sources of noise that are considered totally decorrelated. It concerns in particular the thermomechanical noise S^x_T (NEMS), Johnson noises S^x_J (NEMS and interconnection), 1/*f*-noise $S^x_{1/f}$ (NEMS and interconnection) and electronic noise S^x_E. A_c is the RMS value of the displacement of the NEMS (critical amplitude of a nanobeam), which can be controlled by electric nonlinearity[9] (saturation of the gain electronics or active control of this gain according to the mechanical amplitude):

$$S^x_{openloop}(\omega=\omega_0)=S^x_T(\omega_0)+S^x_J(\omega_0)+S^x_{1/f}(\omega_0)+S^x_E(\omega_0) \qquad [2.51]$$

It must be stressed that the electric noises expressed in V²/Hz are calculated in this instance at the input of the OL system and are expressed in m²/Hz. This conversion depends explicitly on the detection gain K_D. Upon referring to Figure 2.22, to express the spectral density initially expressed in V²/Hz in noise formulated in m²/Hz, its value is divided by the square norm of K_D:

$$S^x_{Electronics}(\omega_0)\left[\mathrm{m}^2/Hz\right]=\frac{S^v_{Electronics}(\omega_0)\left[\mathrm{V}^2/Hz\right]}{|K_D|^2\left[\mathrm{V}^2/m^2\right]} \qquad [2.52]$$

A quick examination of equation [2.50] shows that the strongest possible *SNR* is required. This passes through a large vibration amplitude of the NEMS while remaining in a linear regime. In other words, frequency stability is as good as mechanical work is large.

---

9 The amplitude of oscillations can be controlled using the nonlinear character of the MOS248 or by dynamically adjusting the transconductance gain *gm*, which is a function of the DC drain current *ID*. The first solution is prioritized as it does not require any additional components; the price to pay is the appearance of parasitic harmonics and/or a mix of noise.

Equation [2.52] shows how it is necessary not only to reduce the noise of the NEMS and its electronics as much as possible, but also to optimize the effectiveness of the detection characterized by K_D. Moreover, in order to limit the noise, it also appears essential to have the strongest possible *SNB* in order to obtain an abrupt phase jump (see Figure 2.4(b)), making it easier to put the resonator in oscillation via the self-oscillating phase-locked loop (PLL). In other words, the coupling of actuation and detection must be limited as much as possible by using different measurement strategies or a heterodyning scheme (see Figure 2.27) and using what are known as orthogonal transductions (actuation and detection do not have the same physical origin: for instance, a capacitive actuation and a piezoresistive detection). Equations [2.50]–[2.53] enable an estimation of the resolution of a frequency sensor to be made, regardless of the physical parameter to be measured.

We will now provide details about the main sources of noise mentioned in expression [2.51]. It is, in fact, important to identify and estimate the main noises in the measurement chain in order to optimize *SNR*. Read [CLE 02b] for a complete study. The three main sources of noise are:

– *Johnson noise from resistance*: at a given temperature T, the charge carriers of a resistance R move randomly, generating a random variation of current each time the electrons pass through a given section of resistance. The spectral density of this noise, which is known as Johnson–Nyquist noise, is expressed by the Nyquist relation:

$$S_J(\omega) = 4k_B TR \; [V^2/Hz] \qquad [2.53]$$

where k_B is the Boltzmann constant. In the case studied, R is the resistance of the connections and the resistive parts of the NEMS (e.g. the resistance of the aforementioned nanowires);

– *1/f noise or Flicker noise from a conductor*: the origins are due to different phenomena that are difficult to model. This noise originates from impurities that randomly release or capture charge carriers at electron–hole recombination centers or at the elastic or anelastic distribution of charge carriers by the surface. This last phenomenon may dominate in the case of silicon nanogauges. The resulting noise spectral density follows a law approximately in *1/f*. The empirical model put forward by Hooge [HAR 00] stipulates that spectral density is inversely proportional to the total number of charge carriers contained in the volume of the conductor considered:

$$S_{1/f}(\omega) = \frac{\alpha_H V_{Bias}^2}{Nf} \; [V^2/Hz] \quad [2.54]$$

where α_H is the Hooge constant that depends on the type of conductor between 10^{-4} and 10^{-7};

– *noise from the electronics*: the readout electronics (e.g. LIA) also causes a noise whose spectral density is the superposition of a Johnson noise and a $1/f$ noise originating from active components (CMOS transistors) and interconnection. Nevertheless, above a few kilohertz, the noise in $1/f$ of these components is negligible in comparison to the Johnson term.

– *thermomechanical noise*: its expression can be defined by the fundamental fluctuation–dissipation theorem, which stipulates that any system dissipating energy is noisy. The force exerted by gas molecules around the resonator is responsible for its thermalization. At thermodynamic equilibrium, the fluctuation–dissipation theorem states that this force is the source of dissipation in NEMS. This dissipation originates from the damping term b from equation [2.13]. The quality factor Q is deduced from b according to equation [2.14]. The force noise is a spectral density white noise $S_F(\omega)$ with a centered Gaussian distribution:

$$S_F(\omega_0) = 4k_B Tb = \frac{4k_B T m_{eff}\omega_0}{Q} \; [N^2/Hz] \quad [2.55]$$

The expression of noise spectral density in displacement S_T^x can be determined by equation [2.55] and the frequency response given in [2.15]:

$$\begin{aligned} S_F(\omega) &= \left|\alpha_{NEMS}(\omega)\right|^2 \frac{4k_B T m_{eff}\omega_0}{Q} \\ S_F(\omega_0) &= \frac{4k_B TQ}{m_{eff}\omega_0^3} \end{aligned} \; [m^2/Hz] \quad [2.56]$$

Dominant noises can vary depending on the operating regime of the NEMS (under air, under vacuum, at low temperature, etc.) [EKI 04]. In experiments, the frequency stability of the NEMS in a self-oscillating loop (as shown in Figure 2.29) is often evaluated by measuring a variance in the normalized instantaneous frequency at its nominal value, which is known as Allan deviation (*AV*). This variance is defined by [RUB 05]:

$$AV = \sigma_{\frac{\omega}{\omega_0}}(\tau) = \sqrt{\frac{1}{2(N-1)}\sum_{1}^{N-1}\left(\frac{\bar{\omega}_{i+1} - \bar{\omega}_i}{\omega_0}\right)^2} \qquad [2.57]$$

where ω_i is the average value of angular frequency integrated into the ith duration interval τ and N is the number of independent measures. Equation [2.57] takes into account all the aforementioned sources of noise. This tool enables all the sources of noise to be identified through the different asymptotes (see Figure 2.30). In particular, the noises and long-term derivations in $1/f$ and $1/f^2$ are characterized by a plateau in τ^0 and a gradient in $\tau^{1/2}$, respectively. White noises (thermomechanical and Johnson's) present a gradient in $\tau^{-1/2}$, which comes from the combination of equations [2.49] and [2.50] when only considering white sources of noise. It should be noted that equation [2.57] tends toward true variance (given in equation [2.49]) if the number of samples tends toward the infinite.

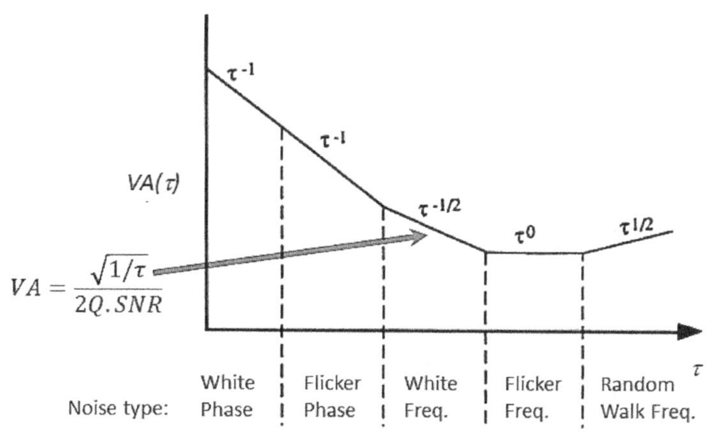

Figure 2.30. *Allan deviation according to integration time τ (over which normalized frequency is averaged). Each asymptote is a characteristic of a type of noise. It will be noted that white noises (e.g. thermomechanical, Johnson's) have a gradient in $1/\tau^{-1/2}$ and Flicker noise (1/f) presents a plateau*

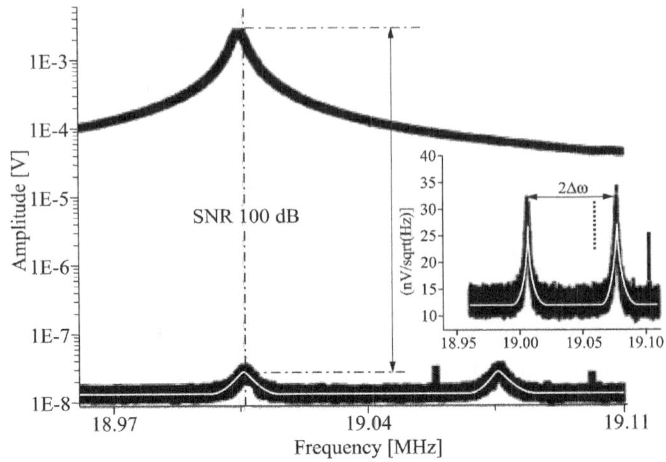

Figure 2.31. *Signal-to-noise ratio (in amplitude) measured on an X-beam (cantilever 5 μm long, 300 nm wide; gauges 500 nm long and 80 nm wide; NEMS 160 nm thick) – NEMS placed under vacuum $P = 10^{-6}$ and $T = 300$ K. The higher curve corresponds to spectral response for an actuation $V_{Bias} = 1.5$ V and $V_{AC} = 1.5$ V (peak). The shift in frequency $\Delta\omega$ is set at 25 kHz. Noise (the lower curve) is measured in the same way without actuation*

Before concluding this chapter, let us return to our two examples of NEMS: the doubly clamped silicon nanowire and the X-beam. We concluded that the simple nanowire presented relatively weak vibration amplitudes, which were limited by mechanical nonlinearities. The X-beam enabled much stronger amplitudes to be reached and had a transduction which, in principle, was more efficient. This is due to an *SBR* which is much better suited to the X-beam. Indeed, it can be seen that *SNR* is very high (100 dB according to Figure 2.31). In this example, the vibrations of the NEMS are measured using a down-mixing measurement scheme with an actuation V_{AC} of 1.5 V (without continuous component). The readout voltages of the gauges are 1.5 V maximum. The size and the principle characteristics of the NEMS are provided in the key. Noise is measured without actuation and therefore corresponds to the sum of the electric noises and thermomechanical noise. For exercise purposes, we can try to evaluate the amplitudes of the different noises. First, the difference between noise on resonance S_{on} and off resonance S_{off} is $S_T^{1/2} = \sqrt{S_{on} - S_{off}} = 17.6 nV / \sqrt{Hz}$.

This value corresponds to the thermomechanical noise caused by the erratic displacement of the beam. Johnson and LIA noises (it is estimated that noise in $1/f$ is negligible in this example, given doping and the volume of nanogauges) are calculated from $S_{off}^{1/2} = \sqrt{S_J + S_V}$. Johnson noise S_J is estimated at $12.8nV/\sqrt{Hz}$ with equation [2.53]. Finally, we can easily deduce the noise of the LIA S_V at $5nV/\sqrt{Hz}$ which is, in fact, coherent with the specifications of the instrument.

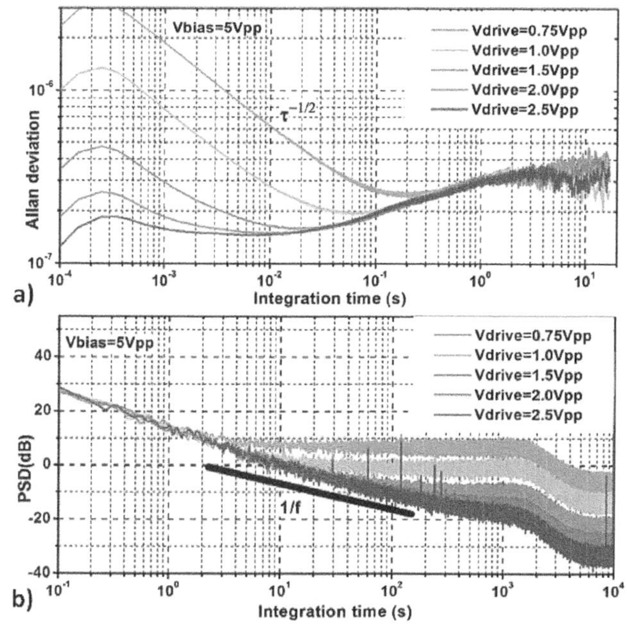

Figure 2.32. *a) Allan deviation of a crossbeam placed under vacuum 10^{-6} T and at room temperature 300 K for different actuation voltages. b) The corresponding noise spectral density measured (calculated from temporal data by realizing a Fourier transform). For a color version of the figure, see www.iste.co.uk/duraffourg/nems.zip*

In addition to *SNR*, frequency stability is usually estimated by *AV* according to the operating points applied. The NEMS is actuated at its resonance frequency and its relative variation is registered according to integration time (acquisition). Measurements can be realized in OL by first estimating the phase variation and then deducing the frequency variation from it (using equation [2.50]). Measurements are also made in closed loop (PLL or self-oscillators). We will not provide details on the different measurement methods used; if interested, refer [VER 06]. Allan deviation and noise spectral density (in dB) for different actuation voltages are shown in Figure 2.32. The NEMS is an X-beam similar to that presented in

Figure 2.26. The Allan deviation in Figure 2.32(a) has a gradient in $\tau^{-1/2}$ for short-term integrations corresponding to the white noises. This tendency is coherent with equations [2.49] and [2.50]. At longer integration times, the Allan deviation presents a plateau which corresponds to the noises in 1/*f*. When actuation voltages increase, amplitude increases, thereby linearly improving *SNR* in the white noise zone[10]. In Figure 2.32(b), white noise corresponds to a plateau that reduces with actuation voltage. At long term (for *AV*) or low frequency (for *PSD*), 1/*f* noise dominates and remains constant regardless of actuation voltage.

2.4. Conclusion

This chapter has presented the different transduction principles (actuation and then detection), estimated their efficiency in terms of *SBR* and studied the difficulty of interfacing with macroscopic measurement instruments. Though scaling laws are *a priori* beneficial for electrostatic actuation and capacitive detection, the parasitic capacitances linked to instrumentation damage this type of detection. Their presence even appears to be unacceptable at high frequencies (>10 MHz). Nevertheless, this constant is valuable for piezoelectric detection. At high frequencies, the strong capacitance values of piezoelectric layers placed in parallel to the input impedances of measurement tools form low-pass circuits which reduce the useful electromechanical signal. Magnetic transduction seems to be very efficient in all circumstances, though it does require strong magnetic fields, thus obliging the researcher to work at a low temperature with a strong volume, even though attempts to integrate the magnet by close fields have been made. Piezoresistive detection remains the ideal detection, though an effort must be made when designing to optimize the constraints in the piezoresistive elements. The different examples also showed that the measurement scheme had a very strong influence on the result. As such, a differential more or less complex heterodyne detection diagram could be used to limit the background. The benefits of the small size become less attractive when the price to pay is a measurement instrumentation that is difficult to implement. To improve and simplify the readout method, a first electronic circuit must be integrated as close as possible, thus isolating the NEMS' output from the macroscopic world[11]. This enables us to return to a

10 When bias voltage is considered, the dominating white noise is thermomechanical noise

11 This approach is the only solution for using capacitive detection.

direct readout without heterodyning and to eliminate cable capacitance effects that reduce the useful signal. For this reason, research activity leading to the cointegration of the NEMS with its CMOS electronics has been developed in recent years. Chapter 3 provides more details on this topic.

Research into cointegration has also led to active NEMS components combining a mechanical part and a transistor, which are slightly different to the SG-MOSFETs described in this chapter. This approach, which will be briefly described in the next chapter (Chapter 3), has notably used scaling effects, such as depletion effects or giant piezoresistive effects. Other transduction methods that use scaling effects have also been created. Chapter 4 will therefore seek to present some examples of scaling effects and their use for detecting the nanometric movement of the NEMS. Optomechanical transduction will also be introduced.

3

Monolithic Integration of NEMS with their Readout Electronics

3.1. Foreword

3.1.1. *Why integrate NEMS with their readout electronics?*

Monolithic integration, also known as "co-integration", designates a scheme integrating nanoelectromechanical systems (NEMS) structures with their electronic readout and processing circuit, which is generally complementary metal-oxide-semiconductor (CMOS)-type, in which these two objects are placed on the same chip[1] and as such are in close proximity to each other. This juxtaposition has considerable advantages, notably the compactness of the system and the unparalleled electrical transduction efficiency.

This spatial proximity enables the effects of parasitic coupling and attenuation to be limited, unlike what is known as the "stand-alone" approach, in which the NEMS and its circuit are placed on separate chips or objects. Limiting these effects consequently maximizes at circuit input both the absolute value of the useful signal and its ratio in comparison to background signal (signal-to-background ratio, SBR). When the circuit is well designed, the latter can measure the intrinsic signal-to-noise ratio (SNR) of the NEMS, an ideal situation in which the system does not perturb the readout of the NEMS, and translates its intrinsic characteristics with regard to noise and detection limit.

---

1 For simplification purposes, in this chapter, the integration of NEMS with their readout electronics will inaccurately be called "NEMS-CMOS integration".

What is more, in contrast to the stand-alone approach which would require an extremely high number of contact pads, co-integration enables NEMS to be deployed in arrays of thousands of units or more and in addition for them to be read individually. In this case, each NEMS is said to be "individually addressable" and is fitted with its own readout circuit or a global circuit, which sequentially examines each NEMS via on-chip multiplexing, depending on the architecture of the system. This advantage appears considerable, given that it is predicted that future applications of NEMS will require arrays to improve redundancy/robustness, the sensing area and the potential to have access to spatial (localized) information. It can be argued, therefore, that a simpler alternative exists from the point of view of the system architecture and practical implementation: the NEMS collective network. Conceptually, this involves a unique input for actuation and a unique output for the detection signal, which is the total of the individual signals. In addition to dispersion problems, this configuration does not enable individual events to be detected: the frequency change of a single NEMS due, for example, to a particle landing on its surface, will not be detectable from the overall signal of the array. For instance, if we consider the mass spectrometry application, the array will be subjected to successive random spatial distribution particle landings and, as such, it will be essential to examine each NEMS separately and increase the number of NEMS such as to boost the probability of sensing a particle. The array can, therefore, be seen as an image constituted of a set of pixels, whose global architecture requires co-integration.

3.1.2. *What are the differences between MEMS-CMOS and NEMS-CMOS?*

The monolithic integration of microelectromechanical systems (MEMS) with CMOS has been a technical, industrial and commercial reality for some time (see [BAL 05, BAL 14, FED 08] for more information). There was a golden age of MEMS-CMOS, but nowadays the ultra-standardization of CMOS production lines and the existence of CMOS foundries which deliver reasonably priced circuits have meant they have almost disappeared from industry. Manufacturers of integrated circuits are apparently very reticent given that both CMOS and MEMS coexist on the same production line. This can, in fact, perturb CMOS yield, reduce production rate and generate yield that is lower overall than that obtained by separate production (monolithic

MEMS-CMOS yield is by definition lower as it is the product of the yields of both). It would, therefore, appear that cost and productivity requirements have led to the pre-eminence of the "stand-alone" approach, i.e. the use of two separate chips to integrate MEMS with the CMOS. Having said that, the race to miniaturize inertial sensors, the large majority of which are capacitive and therefore very sensitive to parasitic coupling and attenuation due to interconnections, could soon change the state of affairs. Miniaturization necessarily reduces the capacitive coupling area and thereby the useful signal and the resulting SNR. Reducing the size of the component while maintaining its basic performance should oblige manufacturers to reconsider certain paradigms related to the integration of MEMS. It is no coincidence that a company like *InvenSense* constantly succeeds in miniaturizing their products while maintaining or even improving their performance. Named after its inventor and founder of the business, the "Nasiri process" consists of co-integrating MEMS and CMOS in a particular way, which is not monolithic in the sense that it uses the same chip, though almost: the CMOS forms the cap of the MEMS and the thus formed cavity is hermetic and under vacuum. The electric connections between them are made via the AlGe eutectic sealing ring.

In addition, MEMS require MEMS-specific process steps and often use active layers with thicknesses that are very different to those commonly used in CMOS processes, idem for certain unusual materials in the world of CMOS (AlN, PZT, SiC, etc). In comparison, NEMS are currently moving toward metal-oxide-semiconductor (MOS) transistors in terms of size (50–500 nm wide and several microns long), thickness (thin silicon on insulator (SOI), for instance, < 200 nm), processes and materials. In other words, unlike MEMS, NEMS can be included in CMOS technology without making any major changes to the process. As we have previously stated, it seems that the key applications for future NEMS require functionalities linked to arrays, which can only be procured by monolithic integration: the future, therefore, looks bright for NEMS-CMOS. Having said that, aside from the problem of production yield, the question of the cost of NEMS has been raised: will one or several application(s) generate enough volume and margins to justify the investments and costs induced by this integration scheme? Only time will tell.

This chapter has three parts. The first part (section 3.2) will (1) describe the advantages of the monolithic integration of NEMS with CMOS in terms of electrical performance, (2) discuss how closed-loop NEMS-CMOS are the essential building block for future NEMS-based frequency sensors and (3) provide an overview of the main achievements on the topic. The second part (section 3.3) will, for education purposes, describe the operating principles of several already published research projects from the perspective of electromechanical transduction. Finally, the third part (section 3.4) will briefly list the future perspectives for NEMS-CMOS in terms of architecture and manufacturing technology.

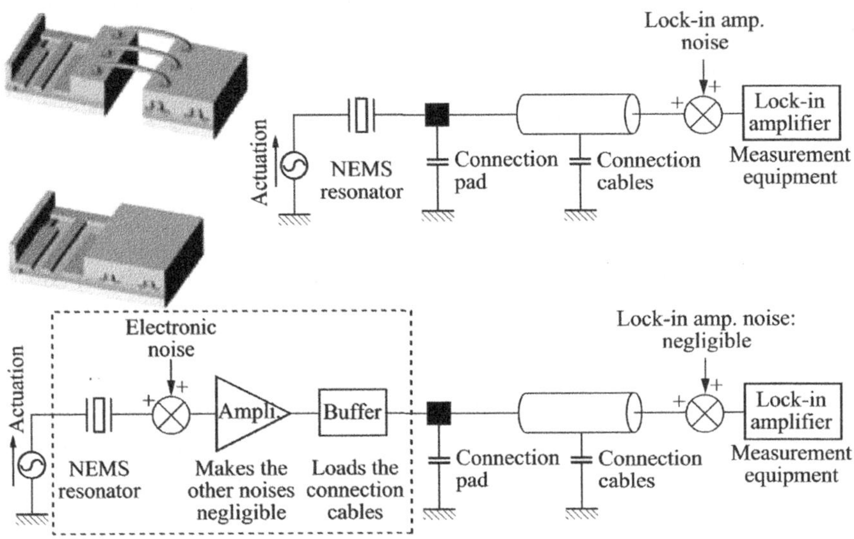

Figure 3.1. *Electrical operating schemes: a) of a "stand-alone" NEMS resonator connected to its external readout electronics. The parasitic capacitance at the NEMS output (the pad and the cables producing capacitances of pF and tens of pF, respectively) generates considerable loss of signal; b) of a monolithically integrated NEMS resonator with its readout circuit that adapts the impedance of the resonator and enables these same output capacitances to be loaded without loss of signal*

3.2. The advantages of and main approaches to monolithic integration

3.2.1. *A comparison of integration schemes and their electrical performance*

The main reason why co-integration considerably improves the performance of electrical transduction in comparison to the stand-alone approach is that it appropriately adapts the impedance of the NEMS to its external environment (Figure 3.1). Indeed, the large majority of NEMS are high-impedance devices and, as a result, present major impedance disadaptation with external interconnections and devices, especially since frequency is high. In the stand-alone approach, NEMS are directly linked to output capacitances (Figure 3.1(a)) which come from the contact pad and the external cables. Their equivalent impedance is often much lower than that of NEMS (Z_{NEMS}), which generates a low-pass filter (otherwise known as a voltage divider bridge) which strongly reduces the useful signal.

We will use a few examples to illustrate this phenomenon: let us consider the typical values of the pad capacitance C_{IN_PAD}=1 pF and the cable capacitance C_{IN_CABLE}=10 pF, and operating frequencies f of 10 and 100 MHz. Let us also consider a typical value of $C_{IN\_NEMS\text{-}CMOS}$=50 fF as a parasitic link capacitance between a NEMS and its circuit in a monolithic configuration. The equivalent impedance of these capacitances $Z_{EQ-C} = 1/(2\pi f C_{IN})$ is shown in Table 3.1.

	f (MHz)	
C	10	100
C_{IN_PAD}=1pF	$Z_{EQ\text{-}C}$=16 kΩ	$Z_{EQ\text{-}C}$=1.6 kΩ
C_{IN_CABLE}=10pF	$Z_{EQ\text{-}C}$=1.6 kΩ	$Z_{EQ\text{-}C}$=160 Ω
$C_{IN\_NEMS\text{-}CMOS}$=10fF	$Z_{EQ\text{-}C}$=1.6 MΩ	$Z_{EQ\text{-}C}$=160 kΩ

Table 3.1. *Equivalent impedance values of different connection capacitances between the NEMS and its readout circuit for different operating frequencies*

In addition, let us, for general cases, consider the loading impedance Z_L seen by the NEMS (and neglecting the series resistance in first approximation). Z_{IN} is the input impedance of the readout electronics and C_{IN} is the parasitic link capacitance (see the three values defined above):

$$Z_L = \frac{Z_{IN}}{1+j2\pi Z_{IN} C_{IN} f} \quad [3.1]$$

In the stand-alone approach, $C_{IN} \approx C_{IN\_PAD} + C_{IN\_CABLE}$. What is more, whatever the value of Z_{IN}, which is either 50 Ω or very high (external measurement equipment generally offers two options), Z_L is generally much lower than Z_{NEMS} (see Table 3.1), which consequently considerably lowers the signal of the NEMS.

On the other hand, in a monolithic configuration, Z_{IN} is generally very high, therefore, Z_L takes the impedance value equivalent to $C_{IN\_NEMS\text{-}CMOS}$, i.e. a value comparable or higher than Z_{NEMS}. In other words, the signal of the NEMS will only be reduced a bit, or not at all. These qualitative considerations will be completed with examples detailed in section 3.3.

With regard to circuit architecture, the literature provides several typologies with several common denominators: circuits generally consist of an amplification stage (generally in voltage) with high input impedance (see the previous sections), then a buffer (unit-gain) enabling impedance matching between the output of the amplifier and the external connections (the capacitances of the pads and wires, equivalent to several pF). It is important to remember that even with a global unit-gain between the NEMS output (circuit input) and the buffer output, the circuit fulfills its mission of impedance matching and allowing the signal to be read out in favorable conditions. By adding further classic-design amplification stages (though at the cost of size and consumption), the circuit can provide a substantial voltage gain, which may enable it to meet the Barkhausen conditions (Chapter 2, section 2.3) and thereby implement a self-oscillating NEMS-CMOS.

Let us finalize these general comments by specifying that piezoresistive (PZR) NEMS very generally present impedance that is much lower (by several orders of magnitude) than that of capacitive NEMS. In comparison to the latter, they are therefore much less sensitive to feedthrough parasitics, which degrade the signal-to-background ratio (SBR) as well as the parasitic output capacitance C_{IN} which reduces the signal (and the SNR of the sensor, see below). Colinet *et al.* [COL 09a] present a general comparison of these two NEMS-types and tend to show, for the aforementioned reasons, that PZR NEMS are more favorable when a frequency of 10–20 MHz is exceeded.

The output signal (V_{OUT}) to noise (V_{NOISE}) ratio, denoted as SNR_{SENSOR}, for a given bandwidth (BW) of a sensor, is as follows:

$$SNR_{SENSOR} = \sqrt{\frac{V_{OUT}^2}{\int^{BW} V_{NOISE}^2 df}} \qquad [3.2]$$

The Robbins formula [ROB 83] mathematically translates the link existing between the (electrical) SNR of the sensor and its phase noise (to which noise is linked the frequency noise as will be shown below):

$$S_{\varphi\_SENSOR} = 1/SNR_{SENSOR} \qquad [3.3]$$

In other words, designing a sensor sensitive to frequency changes, such as NEMS, which has the lowest possible phase noise (or frequency noise) necessarily results in the optimization of the electrical SNR and thereby transduction efficiency.

The fundamental SNR limit of the sensor is set by the SNR of its sensitive part. In the best case, SNR_{SENSOR} cannot, therefore, exceed SNR_{NEMS}. In a stand-alone configuration, given that the signals coming from the NEMS (resonance, background and noise signals) are generally very low at the readout electronics input, the noise of the readout electronics referred to its input, therefore, generally exceeds the noise of the NEMS. In this case, the measured SNR is the ratio between the NEMS signal and the noise of the electronics and is therefore lower than SNR_{NEMS}. Monolithic integration, which reduces NEMS signals a little or not at all, presumably provides the most favorable situation such that the SNR of the sensor reaches its maximum value, i.e. $SNR_{SENSOR} = SNR_{NEMS}$. The diagram in Figure 3.2 compares these two configurations.

Let us consider when $SNR_{SENSOR} = SNR_{NEMS}$ (see the right-hand side of Figure 3.2). The phase noise of the sensor S_{φ_SENSOR} is, therefore, equal to that of the NEMS (see equation [3.3]) S_{φ_NEMS}. The latter is linked to frequency noise through the following relation:

$$S_{f\_SENSOR} = S_{f\_NEMS} = \frac{\partial f}{\partial \varphi} S_{\varphi\_NEMS} \approx \frac{f_0}{2Q} \frac{1}{SNR_{NEMS}} \qquad [3.4]$$

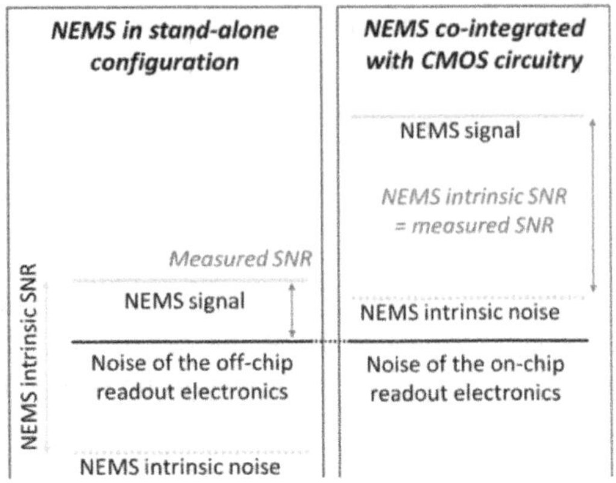

Figure 3.2. *Diagram (for a given bandwidth) of the improvement in SNR procured by monolithic integration. In this case, the complete NEMS+CMOS system reaches maximum SNR, i.e. the intrinsic SNR of the NEMS*

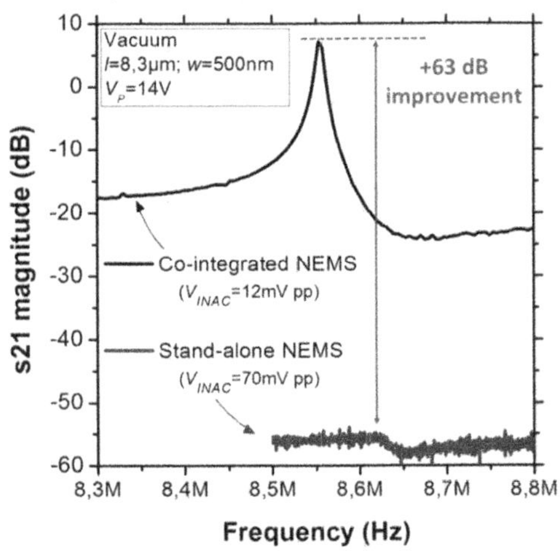

Figure 3.3. *Experimental demonstration of the improvement in SBR and SNR due to monolithic integration. This graph [ARC 12] compares the electrical responses of two morphologically identical resonators, manufactured with the same technology and actuated with the same operating points, one is "stand-alone" and the other is co-integrated (reminder: $s21=20\log[V_{OUT}/V_{INAC}]$). For a color version of the figure, see www.iste.co.uk/duraffourg/nems.zip*

The underlined term only occurs if $\partial f/\partial\varphi = f_0/2Q$ (max value of $\partial f/\partial\varphi$), a condition that is only met when the phase shift before and after resonance is close to 180°, that is to say when SBR is high. In the opposite case, $\partial f/\partial \varphi$ is lower, thus degrading frequency noise. SBR is, therefore, an important point to consider when designing the sensor. In this regard, monolithic integration also procures an improvement. In fact, the spatial proximity between the NEMS and its circuit (which amplifies its signal) enables the feedthrough parasitics found in stand-alone devices to be limited. It is complicated to quantitatively model this feedthrough parasitic coupling: it principally comes from the fringing-field effects of the electric field between the input and output electrodes of the NEMS, electromagnetic coupling between bonding wires, couplings between electrodes and/or routings via the substrate (of capacitive and resistive nature). Countless experimental measurements of different configurations (different directions of routings and bonding wires, shape of the electrodes, NEMS-specific cabling, etc.) corroborate the presence of this feedthrough and its significant reduction by monolithic integration. Figure 3.3 [ARC 12] is an excellent example: it compares the electrical responses of two identical resonators subjected to the same actuation, one stand-alone, the other co-integrated.

3.2.2. *Closed-loop NEMS-CMOS oscillators: the essential building block for NEMS-based frequency sensors*

A sensor using a NEMS as a sensitive element, i.e. that detects resonance frequency changes induced by physical or chemical stimuli, must be capable of detecting both short and long duration events in real time. It is, therefore, essential for the NEMS to operate in closed loop (see Chapter 2, section 2.3).

There are two principle modes of realizing a closed loop oscillator using a NEMS as an oscillating element, both of which are represented in Figure 3.4: self-oscillators and phase locked loops (PLLs).

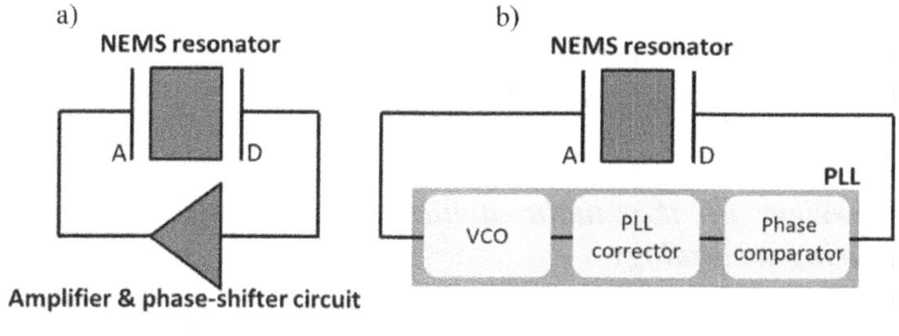

Figure 3.4. *Operating schemes of the two main NEMS-based closed loop oscillator architectures: a) self-oscillator; b) phase locked loop. Letters A and D show the actuation and detection electrodes*

To enter in oscillation and for the oscillations to be self-sustained, self-oscillators must meet the two Barkhausen criteria (see equation [2.47]). The circuit is designed in such a way as to procure gain and phase shift complementary to those of the NEMS and thereby meet the aforementioned criteria. The NEMS-CMOS (co-integrated) self-oscillator is the most compact and smart solution for implementing this solution: as co-integration ensures that the gain of the NEMS is relatively high (though lower than 1), the gain required from the circuit is moderate, which reduces its size and complexity. To date, only two groups (that of N. Barniol at the Autonomous University of Barcelona (AUB) and the NEMS laboratory at the laboratory of electronics and information technologies of CEA (CEA-LETT)) have experimentally demonstrated this type of device. For examples, see [VER 08, VER 13, PHI 14]. Verd *et al.* [VER 13] show a 11 MHz self-oscillator with a (capacitive) NEMS polarization voltage that is lower than the supply voltage of the CMOS (3.3 V), which is a major result. Philippe *et al.* [PHI 14] show a 7.5 MHz self-oscillator with an ultra-compact complete circuit (amplifier + buffer) of seven transistors only with an area of 50 $\times$ 70 μm^2. We will discuss these two examples in more detail in section 3.3.1. In the future, the real-time detection of thousands of individually addressable NEMS in an array will probably be based on thousands of pixels, each one consisting of a NEMS-CMOS self-oscillator (see, for example, Figure18 in [BAT 12]).

A significant number of research projects have focused on external PLL-based closed loop NEMS oscillators [EKI 04a, ARC 11, JOU 13, BAR 12d], but there is no example, in our knowledge, of monolithic association of NEMS with a PLL on a same chip. Implementing an external PLL is relatively accessible, whereas a purely CMOS PLL requires large investments, which research groups did not need to make for sensor demonstration. In the future, industrial applications will perhaps use on-chip dedicated PLL – potentially monolithically integrated with NEMS – as this solution is certainly more flexible and modular than self-oscillators whose operating conditions can be very restrictive.

3.2.3. *Overview of the main achievements from the perspective of manufacturing technology*

This section will present an overview (that is by no means exhaustive) of the research on NEMS-CMOS presented in the literature, from the perspective of manufacturing technology (pre-, intra- or post-CMOS integration of NEMS). Without an exception, research on NEMS and not on MEMS will be discussed here. It must be specified that only a very limited number of groups/research labs in the world are working on (monolithic) NEMS-CMOS, thus resulting in the relatively limited bibliography. In our knowledge, all research concerning NEMS manufactured by top-down technology has a common denominator: as far as possible, CMOS processes have only been slightly modified or not modified at all and the NEMS-specific steps are realized before or after the CMOS (post-processing), or to a lesser extent during CMOS.

The post-processing approach consists of realizing NEMS-specific steps once the CMOS process has ended. One of the most active, or perhaps even the most active group in NEMS-CMOS, the group led by N. Barniol at the Autonomous University of Barcelona (Spain), follows this approach, which is *a priori* the most efficient in terms of the number of additional steps and masks. Most of their research works are based on a 0.35 μm CMOS technology (two polySi, four metals) from AMS (see Figure 3.5(a)) [VER 08, VER 13, VER 06a, VER 07a, LOP 09]. Capacitive NEMS are defined either in the final metallic interconnection layer at the back-end of the CMOS (see Figure 3.5(c) and (d)) or in one of the two polySi layers (see Figure 3.5(a) and (b)) nominally used to realize capacitors. In both cases, the CMOS process does not need to be modified. The key point is gaining authorization from the founder to break some rules in the design kit (DK), in particular those related to pad opening. These openings are realized above each NEMS such as to etch the intermetallic dielectric layers until the resonators are reached. The only post-CMOS step, the release of mechanical structures, is achieved by simple chemical etching, without masking. A high-frequency (HF)-based solution etches the SiO_2 under and over the edges of the structure while being selective with regard to the metal of the pads. The polySi layer-based approach, which is usually used to implement capacitors (see [ZAL 10, ARC 08] for information about this principle), procures an efficient capacitive transduction, due to a 40 nm gap, which is the thickness of the capacitors dielectrics.

As shown in Figure 3.5(b), the NEMS and its electrodes are, therefore, realized in the bottom and top polySi layers, respectively. The AUB's post-processing approach, regardless of the material of the NEMS, has now been proven: its performance, repeatability and ease of implementation have been confirmed in a set of publications. There are, however, disadvantages to this approach, including the relatively high thickness (from 300 nm to 1 μm) of the NEMS and the nature of the materials that are only favorable to capacitive detection and not at all, or very little to PZR detection.

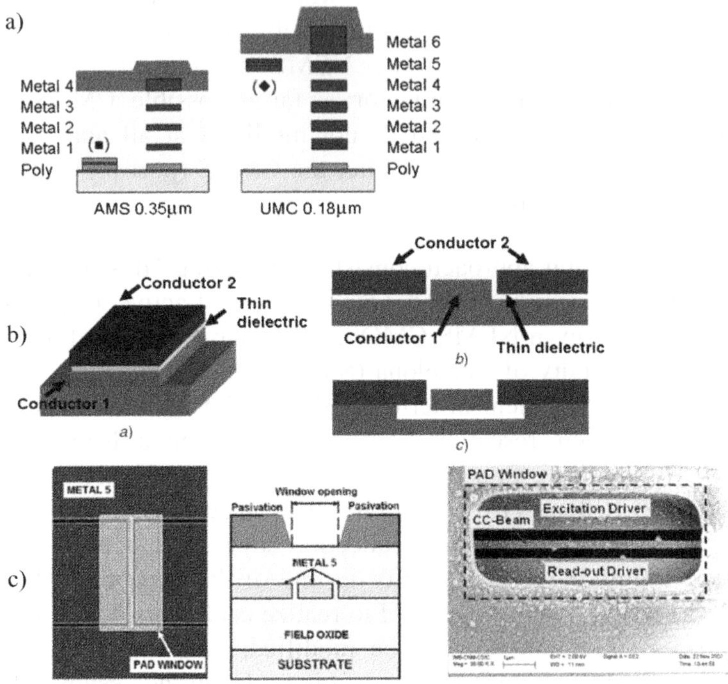

Figure 3.5. *Diagrams taken from [LOP 09] summarizing the NEMS-CMOS post-processing approach of N. Barniol's group at the AUB; a) the CMOS technologies most commonly used by this group: AMS 0.35 μm (NEMS manufactured in M4 or polySi layers) and UMC 0.18 μm (use of M5); b) cross-section of an operating scheme of resonators based on AMS polySi layers; c) layout and cross-section of a resonator using the M5 of UMC; d) SEM view (from above) of a NEMS-CMOS according to the principle described in c). For a color version of the figure, see www.iste.co.uk/duraffourg/nems.zip*

Other relatively recent achievements, which are more MEMS-CMOS oriented, use the set of the back-end stack (groups led by W. Fang and S.-S. Li at the National Tsing Hua University in Taiwan [TSA 12, LI 13]), or a metal deposited by electroplating above the integrated circuit (IC) [HUA 08] (group led by C.T.-C. Nguyen in Berkeley) as the structural layer of the MEMS.

The pre-CMOS approach consists of realizing NEMS in the front-end before the CMOS process begins. One of the most active groups in

NEMS-CMOS, the NEMS lab at the CEA-LETI, follows this approach, which enables monocrystalline Si (Si-mono) to be used as the structural layer of the NEMS. It does, however, require the CMOS foundry to agree to realize a CMOS process on wafers that are not bare Si substrates. The Si-mono combines good mechanical properties (high stiffness – E=169 GPa – and wide elastic range) with the possibility of realizing capacitive or piezoresistive NEMS.

In [ARC 12] and [PHI 14] (see Figure 3.6), manufacturing starts by defining capacitive NEMS in the 1 µm-thick top Si layer of the SOI substrate. The NEMS/electrode gaps, as well as the trenches around the NEMS and the circuit are then filled in with SiO_2. In this example, the foundry (STMicroelectronics) agreed to implement (but with no other modifications) its 0.35 µm bulk CMOS technology on SOI substrate, which facilitates the further release of the NEMS by removing the buried oxide (BOX), which is used as a sacrificial layer. Once the CMOS has been manufactured, the opening of the passivation above the pads also enables the removal of this passivation above the NEMS. Following the deposition/lithography/etching of a protective layer on the circuit, the NEMS are released in post-CMOS by an HF-based chemical etching.

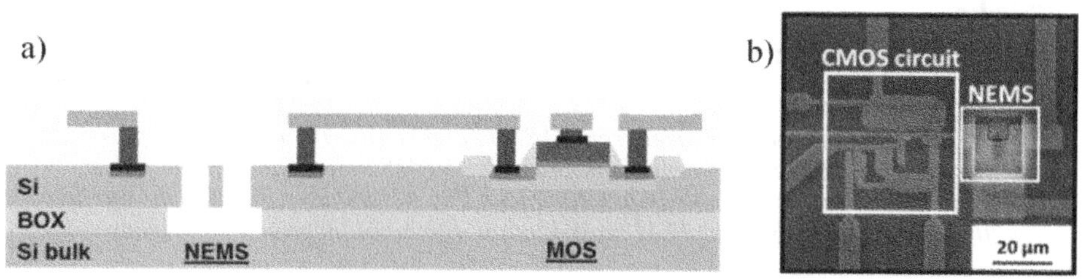

Figure 3.6. *Schemes taken from [ARC 12] and [PHI 14] of the pre-CMOS integration of capacitive NEMS in a 0.35 µm bulk CMOS technology of STMicroelectronics, implemented on an SOI substrate. The NEMS are defined in the top layer (1 µm-thick) of the SOI. For a color version of the figure, see www.iste.co.uk/duraffourg/nems.zip*

Research by CEA-LETI [ARN 12, OLL 12, ARC 14] on the integration of very high frequency (100 MHz) piezoresistive NEMS in Si-mono, with a CMOS fully-depleted SOI (FDSOI) technology (see Figure 3.7), also fall into the pre-CMOS category, though in this case it was necessary to make substantial modifications to the CMOS process: for instance, at the start of the CMOS to compensate for the non-planarity of the wafers surface after the NEMS have been defined, or by providing specific doping. This was only possible by making the full process in-house (at LETI), thus not requiring any foundry to intervene. It should be noted that NEMS are manufactured in the same layer as the FDSOI transistors channel. One of the originalities of this process is its strategy for the NEMS release: once the NEMS have been defined and then released by an HF chemical etching (without a protective layer), they are thermally oxidized and then encapsulated with polySi, all of which in pre-CMOS. At the end of the CMOS process and the opening of the passivation above the NEMS, the latter are released again by selective gas-phase chemical etching (based on CF_4), removing the polySi while leaving the passivation dielectrics intact. In our knowledge, this is the only research to have demonstrated NEM-CMOS with PZR detection, as well as NEMS manufactured by top-down technology with such small sizes (beam: 100 nm wide, 1.2 μm long; PZR gauges: 30 nm wide, 100 nm long, beams and gauges being 20 nm-thick).

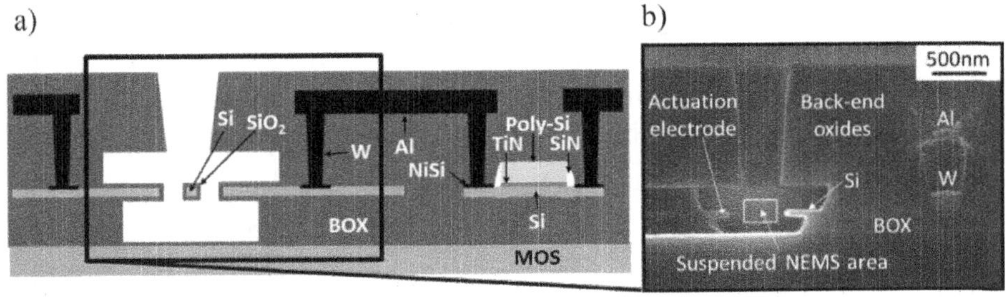

Figure 3.7. *a) Schemes taken from [ARN 12] and [ARC 14] showing the pre- and intra-CMOS integration of piezoresistive NEMS in a 0.3 μm CMOS fully-depleted SOI (FDSOI) technology of CEA-LETI. The NEMS are defined in the top layer (20 nm thick) of the SOI; b) TEM cross-section with a co-integrated MOS transistor*

The general trend for research on NEMS-CMOS is the miniaturization of the NEMS resonator, as illustrated in Figure 3.8. The group led by N. Barniol conducts experiments on technologies such as united microelectronics corporation (UMC) 0.18 μm or STMicroelectronics 65 nm [MUÑ 13]. The CEA-LETI carried out previous research [ARN 12, OLL 12, ARC 14] based on an FDSOI technology that is currently in production at the 28 nm node. The NEMS and the transistor channels are realized in the same 20 nm Si-mono layer. Unlike MEMS-CMOS, NEMS are really converging toward the size of the transistors and their manufacturing processes.

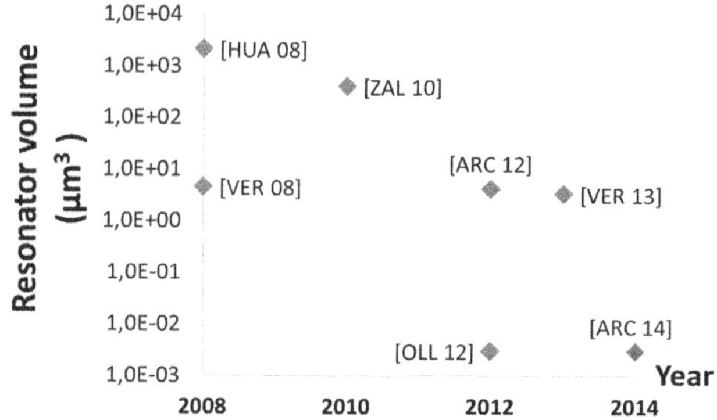

Figure 3.8. *Year-by-year comparison of recent research published on NEMS-CMOS according to the resonator volume*

3.3. Analysis of some significant achievements from the perspective of transduction

The aim of this section is to analyze in more detail, for educational purposes, a number of major examples from the literature on NEMS-CMOS from the perspective of transduction principles (see Chapter 2, section 2.2).

3.3.1. *Examples of capacitive NEMS-CMOS*

3.3.1.1. *NEMS-based 8 MHz self-oscillator in Si-mono (research by CEA-LETI)*

These monolithic components [ARC 12, PHI 14] are based on 1 µm-thick capacitive NEMS realized in the top layer in Si monocrystalline (Si-top) of an SOI substrate, over which a 0.35 µm bulk CMOS technology of STMicroelectronics is implemented (the Si-top is thick enough to ensure the behavior of the transistors will not be modified in comparison to a bulk substrate). The implemented manufacturing technology is described in section 3.2.3. This research constitutes one of the only examples of NEMS-CMOS self-oscillators (alongside [VER 08] and [VER 13]). As previously explained, future sensors sensitive to variations of resonance frequency (f_0), such as high-resolution mass sensors, will likely be based on NEMS arrays; the resonance frequency of each NEMS being monitored individually and in real time. Self-oscillators should, therefore, constitute an essential building block.

3.3.1.1.1. General architecture of the NEMS-CMOS self-oscillator

This system (see Figure 3.9) is based on the oscillation at resonance of a NEMS clamped-free beam (or *cantilever*) in its first mode of lateral bending (in-plane) with the following dimensions: 6 µm long, 250 nm wide, 1 µm thick and with a gap of 250 nm. Its associated CMOS circuit sustains oscillations by providing appropriate gain and phase shift. For the system to oscillate, it must meet the two Barkhausen conditions (see Chapter 2, section 2.3), namely that the total gain G_{TOT} of the system at resonance and in open loop must be higher or equal to 1 and that the total phase shift in open loop must be 0° (2π):

$$|\, G_{TOT}(\omega_0)| \geq 1 \quad \text{and} \quad arg\; G_{TOT}(\omega_0) = 0°(2\pi) \qquad [3.5]$$

where $G_{TOT} = G_{NEMS} \times G_{CMOS\text{-}AMP}$; $G_{NEMS} = V_{OUT\text{-}NEMS} / V_{IN\text{-}NEMS}$; and $G_{CMOS\text{-}AMP} = V_{OUT\text{-}AMP} / V_{OUT\text{-}NEMS}$. Monolithic integration procures a low value of C_{IN}, C_{IN} being the sum of all the grounded capacitances between the NEMS output and the circuit input: $C_{IN}=C_{O2}+C_{PARA}+C_{GS\text{-}M0}$ (see Figure 3.9(c)). C_{O2} is the static capacitance of the detection electrode, C_{PARA} that is between the NEMS-CMOS and the substrate – *de facto* much lower than 50 fF, and $C_{GS\text{-}M0}$ the gate-source capacitance of M0 transistor. G_{NEMS} (see equation [3.6]), therefore, remains potentially very close to (though always lower than) 1. The presence of an amplifier circuit sustaining the oscillations is certainly still necessary, but the gain $G_{CMOS\text{-}AMP}$ to be provided is consequently moderate (between 1 and 10):

$$G_{NEMS} = \frac{V_{OUT-NEMS}}{V_{IN-NEMS}} = \frac{1}{1+jR_M C_{IN}\omega} \qquad [3.6]$$

with (for the first lateral bending mode in the plane of a cantilever):

$$R_M = \frac{\sqrt{k_{EFF} m_{EFF}}}{Q\eta^2} = 0.648\frac{w^2 g^4}{L^3 h}\frac{\sqrt{E\rho}}{QV_P^2\varepsilon_0^2} \qquad [3.7]$$

3.3.1.1.2. Architecture of the CMOS circuit

The oscillation conditions of the NEMS/CMOS system (for example, see [COL 09a]) have determined the circuit design rules (see also [ARN 11]) whose primary goal is to amplify and phase shift the output voltage of the NEMS (node $V_{IN\text{-}CMOS} = V_{OUT\text{-}NEMS}$). This circuit has a simple architecture (see Figure 3.9(c)) based on a single amplification transistor M0 with the required $G_{CMOS\text{-}AMP}$, and a buffer adjusting the output impedance of the latter to that of the external connections. The low complexity of the circuit due to its reduced number of transistors (just seven) results in a very low consumption of energy (about 1.2 mW) and area (ultra-compact NEMS+ circuit device, 50 × 70 μm^2).

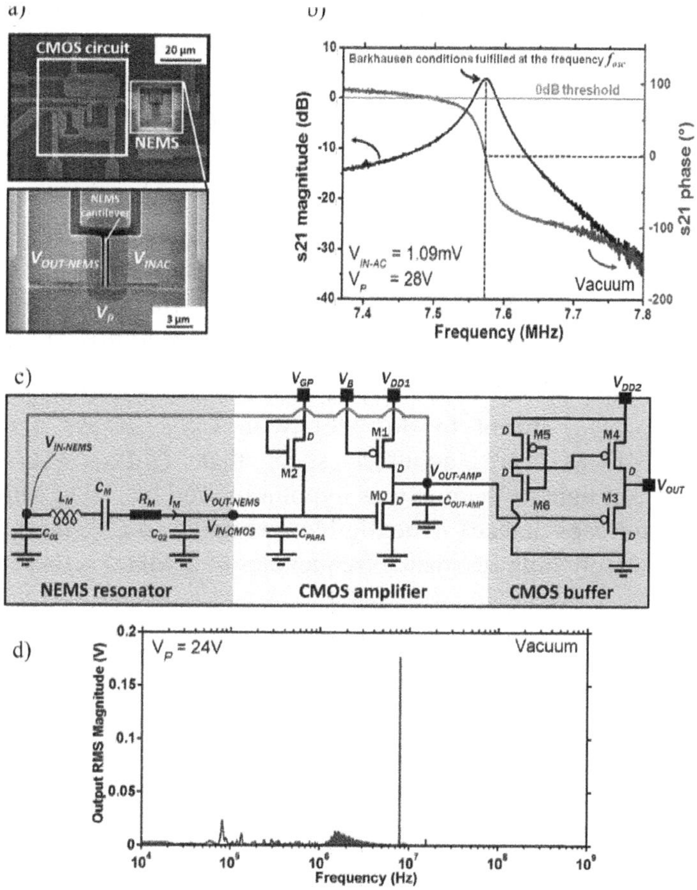

Figure 3.9. *Figures taken and inspired from [ARC 12] and [PHI 14]: 8 MHz NEMS-CMOS self-oscillator based on capacitive NEMS and a 0.35 μm bulk CMOS technology of STMicroelectronics; a) SEM image of a NEMS-CMOS cell (total area of 70 μm²). Zoom (below) on the double electrode configuration of the NEMS (one for AC actuation, the other for detection, the resonator being DC polarized to V_P); b) electrical response in open loop, under vacuum, meeting the Barkhausen conditions; c) equivalent electrical schematic of the NEMS-CMOS in closed loop (the blue wire links the amplifier output to the NEMS input); d) demonstration of self-oscillation (under vacuum) for V_P=24 V: the FFT of the electrical response shows a single peak corresponding to the NEMS resonance (no parasitic oscillations). For a color version of the figure, see www.iste.co.uk/duraffourg/nems.zip*

The capacitive current i_M, known as motional current, produced by the NEMS, is converted into a voltage $V_{IN\text{-}CMOS}$ via the integration capacitance C_{IN}; $V_{IN\text{-}CMOS}$, therefore, is being amplified by M0. The three transistors of the amplification stage play the following roles:

– M1 is a p-type mos (PMOS) transistor polarized on its gate by a DC voltage V_B guaranteeing its saturation. It acts as a current generator with a load resistance R_{LOAD} ($\sim 1/gds_{M1}$). M1 imposes a DC polarization current I_{DS} on the M0 transistor, thereby enabling it to be saturated by setting the DC level of $V_{OUT\text{-}AMP}$.

– M2 is a n-type mos (NMOS) transistor polarized in active load by a DC voltage V_{GP}. Its function is to polarize in DC the M0 gate to a value V_{GP} without causing leakage from the NEMS current toward this transistor. M2, therefore, serves as polarization resistance R_{POL} (high impedance). The dimensions of M2 are such that $(\omega_0 C_{IN})^{-1} \ll R_{POL} \approx (gm_{M2} + gds_{M2})^{-1}$.

– The function of M0 is to convert $V_{IN\text{-}CMOS}$ into a current of intensity ($-gm$ x $V_{IN\text{-}CMOS}$) meant to flow solely in $C_{OUT\text{-}AMP}$. As a result, the dimensions of M1 are adjusted such that $(gds_{M0} + 1/R_{LOAD}) \ll \omega_0 C_{OUT-AMP}$. In light of the previous hypotheses and by sizing M0 such that C_{GD} (gate-drain capacitance) is negligible compared to C_{IN}, the gain of the circuit for relatively high resonance frequencies (> 1 MHz) is therefore given by:

$$G_{CMOS-\mathrm{AMP}} \approx gm_{M0} / (C_{OUT-\mathrm{AMP}}\omega_0) \quad [3.8]$$

It should be noted that the specific DC polarization points (V_B and V_{GP}) have been determined such as to generate the same gate and drain DC voltage over M0, meaning that in a self-oscillator configuration, the NEMS is symmetrically polarized at the same DC potential over its two lateral electrodes (see Figure 3.9(a)), which is preferable for its dynamic behavior. This also ensures that M0 is in saturation regime.

The buffer consists of transistors M3, M4, M5 and M6. Its function is to provide impedance matching, enabling large output capacitances (linked to the measurement instrumentation: contact pads, cables, oscilloscope, etc.) to be loaded, typically 10–20 pF, and therefore to provide enough current. Ideally, the gain of the buffer $G_B = V_{OUT} / V_{OUT\text{-}AMP}$ must be equal to 1, but in practice it is slightly lower. The monolithic integration of this buffer is fundamental for the performance of the circuit: in fact, from the perspective of increasing $G_{CMOS\text{-}AMP}$, the buffer enables values of $C_{OUT\text{-}AMP}$ to remain relatively low (<< pF). If the buffer was external (for example, on an off-chip circuit), $C_{OUT\text{-}AMP}$ would be at least 100 times higher, which would be prohibitive to obtain a sufficient $G_{CMOS\text{-}AMP}$.

3.3.1.1.3. Self-oscillating the NEMS-CMOS system

In light of the NEMS and circuit designs presented in [ARC 12] and [PHI 14], respectively, applying equations [3.6]–[3.8] shows that the gain criterion from the Barkhausen conditions (equation [3.5]) can be met from a minimum polarization voltage V_P of the NEMS of around 20 V. To meet the second criterion, the NEMS-CMOS system is designed such as to generate successive phase shifts, thus enabling a total phase shift of 0° to be obtained:

– at resonance, the phase shift between $V_{IN\text{-}NEMS}$ and i_M is 0°;

– between i_M and V_G , the capacitance C_{IN} generates a phase shift $\Delta\varphi_1$ of –90°;

– between $V_{IN\text{-}CMOS}$ and the amplified intensity current ($-gm\ V_{IN\text{-}CMOS}$) (see Figure 3.9(c)), the phase shift $\Delta\varphi_2$ is –180°;

– between the current ($-gm\ V_{IN\text{-}CMOS}$) and $V_{OUT\text{-}AMP}$, the capacitance $C_{OUT\text{-}AMP}$ generates phase shift $\Delta\varphi_3$ of –90°. To reach this result, all the current ($-gm\ V_{IN\text{-}CMOS}$) must flow solely through $C_{OUT\text{-}AMP}$, implying that $(gds_{M0} + 1/R_{LOAD}) \ll \omega_0 C_{OUT-AMP}$. More generally, the choice of the value of $C_{OUT\text{-}AMP}$ results in a trade-off between meeting the gain criterion (low $C_{OUT\text{-}AMP}$) and meeting the phase shift criterion (relatively high $C_{OUT\text{-}AMP}$, i.e. of low impedance). The trade-off found here resulted in $C_{OUT\text{-}AMP}$ = 400 fF. Once again, monolithically integrating the buffer is fundamental; if it was not the case, $C_{OUT\text{-}AMP}$ would have been too high (as it is directly determined by the size of the pad and the external connections).

In addition, it is absolutely necessary that when it enters in self-oscillation, the NEMS-CMOS system be designed such as to induce a saturation phenomenon, limiting the mechanical amplitude of oscillation to a value lower than that of the dynamic pull-in [KAC 10] (which would otherwise lead to irreversible stiction and perhaps even the destruction of the beam). For an arbitrary maximum gap limit of 50%, this system controls oscillations below this threshold by adjusting G_{NEMS} with an appropriate value of V_P, and putting to good use the saturation of the circuit itself (and not the mechanical and electrostatic nonlinearities, whose effect would not be sufficient).

That the Barkhausen conditions had been met was checked and adjusted in an open loop (see Figure 3.9(b)). The adjustment parameters were then applied to a closed loop structure (the NEMS has no external actuation pad, its input is linked to the node $V_{OUT\text{-}AMP}$) and the self-oscillations were spontaneously created and stabilized, as is shown in Figure 3.9(d). The latter shows the fast Fourier transform (FFT) of the self-oscillation signal (measured under vacuum at node V_{OUT}) at approximately 8 MHz, obtained for a relatively high V_P of 24 V. The frequency spectrum shows a single peak, corresponding to the resonance frequency of the NEMS, and is not perturbed by parasitic resonances.

3.3.1.2. *NEMS-based 11 MHz self-oscillator in polySi (research by AUB)*

These monolithic components [VER 13], by N. Barniol's group at the AUB, are based on capacitive NEMS that are about 300 nm-thick and realized in one of the two polySi layers (which nominally constitute capacitors) of a 0.35 µm CMOS technology of Austria Microsystems (AMS), see Figure 3.10(a). The two NEMS electrodes (AC actuation and detection) are realized in the second polySi layer and enable a 40 nm gap with the NEMS (a doubly clamped tuning fork, 12.8 µm long and 500 nm wide). The implemented manufacturing technology is described in section 3.2.3 (for example, see [VER 06a, VER 07b, LOP 09]).

This system is a Pierce-type oscillator, whose resonant element is a NEMS tuning fork actuated in its first lateral bending mode in counter-phase. The CMOS circuit sustaining the oscillations (see Figure 3.10(b)) is based on a cascode-type differential voltage amplifier with an area of 54 × 66 µm^2 and reduced power consumption (1.5 mW). A voltage-follower buffer enables the output impedance of the amplifier to be matched to a 50 Ω external connection. The ultra-efficient capacitive detection, due to the 40 nm gap, enables CMOS-compatible NEMS DC polarization voltages V_P, that is to say lower than V_{DD}=3 V, which is unprecedented for a NEMS-CMOS self-oscillator.

This differential configuration drastically reduces the background signal (common mode) linked to the feedthrough parasitics of the NEMS, which improves its gain and phase shift. Two identical NEMS are, therefore, connected to the two differential inputs, one being used as a resonant element of the circuit, the other solely for the subtraction of the parasitic coupling current (except for this parasitic current, capacitive resonators, even actuated, do not deliver motional current if their effective V_P is null). In steady-state regime and under vacuum (this is not specified in the paper [VER 13], but it is probably the case), the self-oscillator delivers (Figure 3.10(c)) a peak-to-peak amplitude signal of 180 mV with a frequency of 11 MHz (f_0 of the NEMS). To date, [VER 13], a revised and improved version of [VER 08], seems to be the most successful achievement with regard to NEMS-CMOS self-oscillators in terms of transduction efficiency (visible from the CMOS-compatible low NEMS V_P), the size of the NEMS and phase noise (noise floor of –110 dBc/Hz).

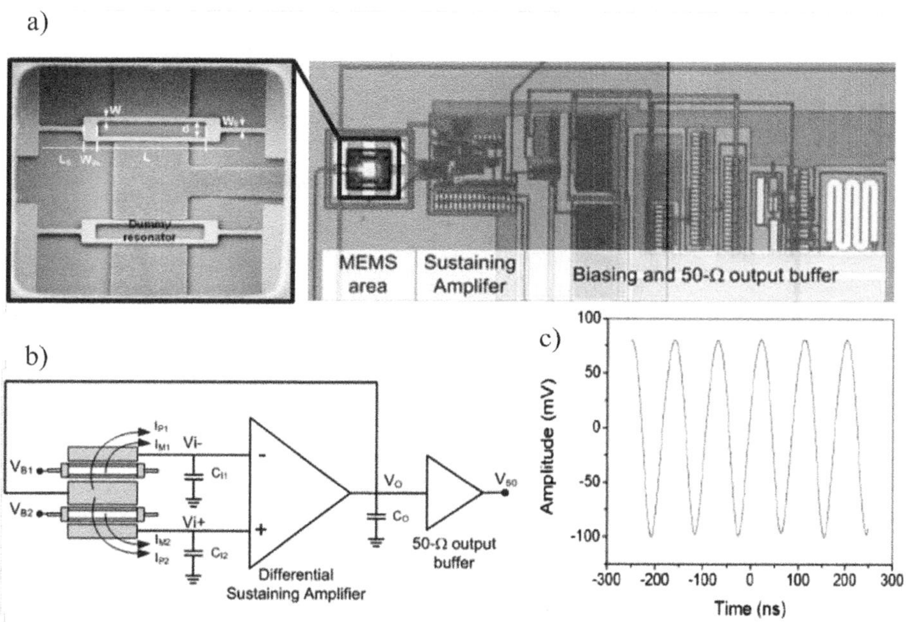

Figure 3.10. *Figures taken from and inspired by [VER 13]: 11 MHz NEMS-CMOS self-oscillator based on capacitive NEMS implemented in an AMS 0.35 µm CMOS technology; a) SEM (left) and optical (right) images of the NEMS-CMOS cell. The NEMS (a doubly clamped tuning fork) has a double electrode configuration: one is for AC actuation and the other for detection, the resonator being DC polarized; b) equivalent electrical schematic of the NEMS-CMOS self-oscillator in Pierce configuration: a differential amplifier is used with two NEMS, one of the two does not deliver resonance signal but enables the parasitic coupling signals to be removed; c) functional demonstration: time-domain output signal at frequency f_0 in steady-state self-oscillation regime*

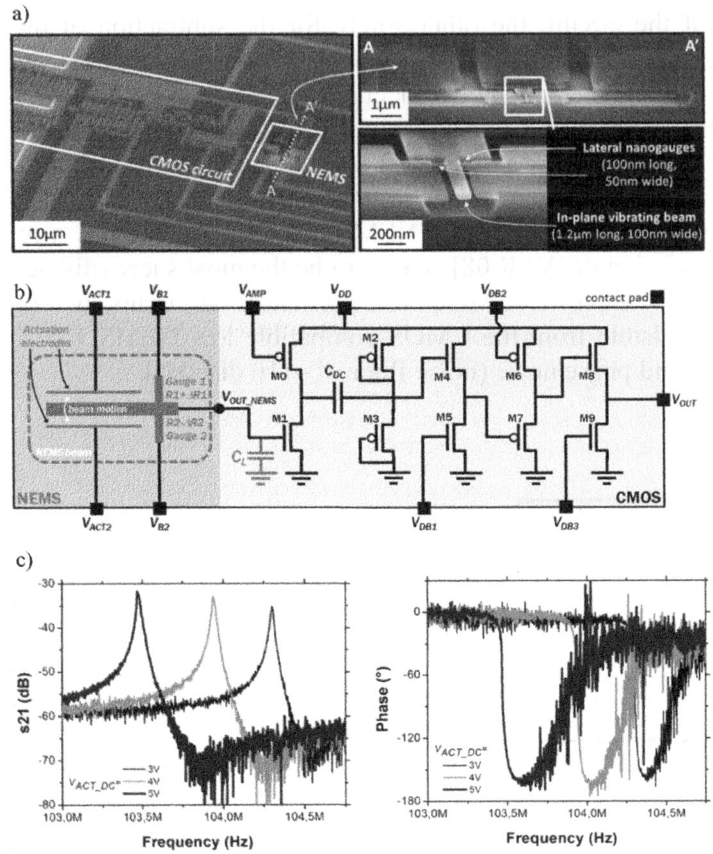

Figure 3.11. *Figures taken from and inspired by [OLL 12] and [ARC 14]: 100 MHz NEMS-CMOS device based on piezoresistive NEMS smallest with ultimate (20 nm thick, 1.2 µm long and 100 nm wide) and sub-100 nm FDSOI-type CMOS technology; a) increasingly magnified SEM images of the NEMS-CMOS cell; b) equivalent electrical schematic of the NEMS-CMOS; c) electrical response (s21 transmission parameter) in terms of amplitude (left) and phase (right) measured under vacuum for a polarization voltage of the gauges (V_{B1}-V_{B2})=4 V and an AC actuation voltage of 2.25 mV rms. For a color version of the figure, see www.iste.co.uk/duraffourg/nems.zip*

3.3.2. *Examples of piezoresistive NEMS-CMOS*

3.3.2.1. *100 MHz "crossbeam" resonators by CEA-LETI with CMOS FDSOI process*

These monolithic components [OLL 12, ARN 12, ARC 14] (see Figure 3.11(a)) are based on piezoresistive NEMS implemented in the top

20 nm-thick Si monocrystalline layer of an SOI substrate. The resonators are "crossbeams" [MIL 10] with ultimate and the smallest yet: the beams are 1.2 μm long, 100 nm wide and have two lateral gauges, 80 nm long and 50 nm wide. These NEMS are realized in the same Si layer as the transistors channels using advanced sub-100 nm FDSOI-type CMOS technology by the CEA-LETI. The complete process for manufacturing this kind of NEMS-CMOS is described in section 3.2.3 (see [OLL 12, ARC 14]).

When published, this research led to significant achievements and novelties: (1) the first demonstration of NEMS-CMOS with PZR detection, (2) some of the smallest, if not the smallest, NEMS manufactured with "top-down" technology and (3) one of the only examples of direct readout (homodyne), and this for a very high frequency (100 MHz). It must first be stated that it is simpler and more compact to implement a homodyne NEMS-CMOS self-oscillator than a heterodyne one. Indeed, as previously explained, the readout of high-frequency NEMS (over 10 MHz), which inherently have very reduced dimensions, generally makes the use of heterodyne detection schemes mandatory for down-converting at low frequency the useful signal which in essence is low. This thereby preserves it from strong attenuations (linked to the low-pass output filter, see equation [3.4] and its related paragraph) and parasitics linked to high frequency. Obtaining a direct readout of such small NEMS at 100 MHz is remarkable. It has been made possible by co-integration which provides a low parasitic capacitance C_{IN} at the NEMS output (here $C_{IN} = C_L + C_{GS\text{-}M1}$, C_L being the parasitic capacitance between the NEMS and the CMOS, and $C_{GS\text{-}M1}$ the gate-source capacitance of the M1 transistor, see Figure 3.11(b)).

The expression of the output voltage of the NEMS ($\equiv V_{IN\text{-}CMOS}$) is as follows:

$$V_{OUT-\mathrm{NEMS}} = \frac{(V_{B2}-V_{B1})}{2+jC_{IN}\omega(2\mathrm{R}_{ba}+\mathrm{R})}\frac{\Delta\mathrm{R}}{R} \approx \frac{(V_{B2}-V_{B1})}{2}\frac{\Delta\mathrm{R}}{R} \quad [3.9]$$

V_{B1} and V_{B2} are the DC polarization voltages of the gauges, ω is the angular frequency of the beam mechanical motion, R_{ba} is the series resistance at the NEMS output (between the common extremity of the two gauges and the M1 gate), R is the resistance of a gauge (identical for both) and ΔR is their variation while they are subjected to a mechanical effort linked to the resonance of the crossbeam in its first in-plane bending mode.

The equality of the bottom is only valid because C_{IN} is low (several tens of fF). The relative variation $\Delta R/R$ is then:

$$\frac{\Delta R}{R} = \frac{G}{E} \frac{\alpha \, Q \, F_{ACT}}{A_J} \qquad [3.10]$$

G is the gauge factor in the direction <110>, E is the Young's modulus, α is a dimensionless factor translating the lever arm effect between the equivalent application point of the actuation electrostatic force F_{ACT} (applied to each of the two electrodes) and the common extremity of the two gauges, Q is the quality factor of the resonance and A_J is the section of the gauges. The electrical response by (homodyne) direct readout, expressed in terms of $s21 = 20log(V_{OUT}/V_{ACT})$, is shown in Figure 3.11(c). These resonance peaks display excellent SBR and SNRs for moderate polarization voltages of gauges (V_{BI}-V_{B2}=4V) and ultra-low AC actuation voltages (2.25 mV rms), thus illustrating the efficiency of electromechanical transduction, notably due to co-integration. It should be noted that to obtain this result, a differential actuation (V_{ACT1} and V_{ACT2} in counter-phase) is absolutely necessary so that the feedthrough parasitic signals can be subtracted on the $V_{OUT\text{-}NEMS}$ node. The Allan deviation measurements show an optimum of $\sigma_A = 2 \times 10^{-7}$ for $\tau = 500$ ms. This corresponds to a mass detection limit as low as 2 zg, i.e. ~1.3 kDa (a value that still needs to be demonstrated in mass measurements experiments).

3.3.2.2. *Resonant body transistors by the Massachusetts Institute of Technology (MIT) with CMOS SOI process of International Business Machines corporation (IBM)*

This research [MAR 14, MAR 12] concerns 11 GHz acoustic RF resonators implemented in resonant body transistors (RBTs) with electrostatic actuation (see Figure 3.12(a)).

Electromechanical detection is based on the PZR modulation of the drain current of an n-type field effect transistor (nFET), the PZR effect inducing a modulation of the carriers mobility. Unlike other research works previously presented in section 3.3, these resonators (with a size of 360 nm × 2.5 µm for a total device size of 5 × 3 µm²) do not contain a CMOS readout circuit; they are destined to realize acoustic frequency sources with reduced size, consumption and parasitics. But they constitute an example of co-integration as they are manufactured at the transistor level of a 32 nm CMOS SOI technology of IBM, using a channel-oxide-gate (in polySi) stack (see Figure 3.12(b)). Integration is facilitated by the fact that these resonators are not released but buried, the lateral acoustic vibrations being confined by the adjacent Bragg mirrors. This research constitutes a first promising step and provides a preliminary demonstration of functionality with a frequency measurement of 11.5 GHz ($Q \sim 24$) (see Figure 3.12(c)).

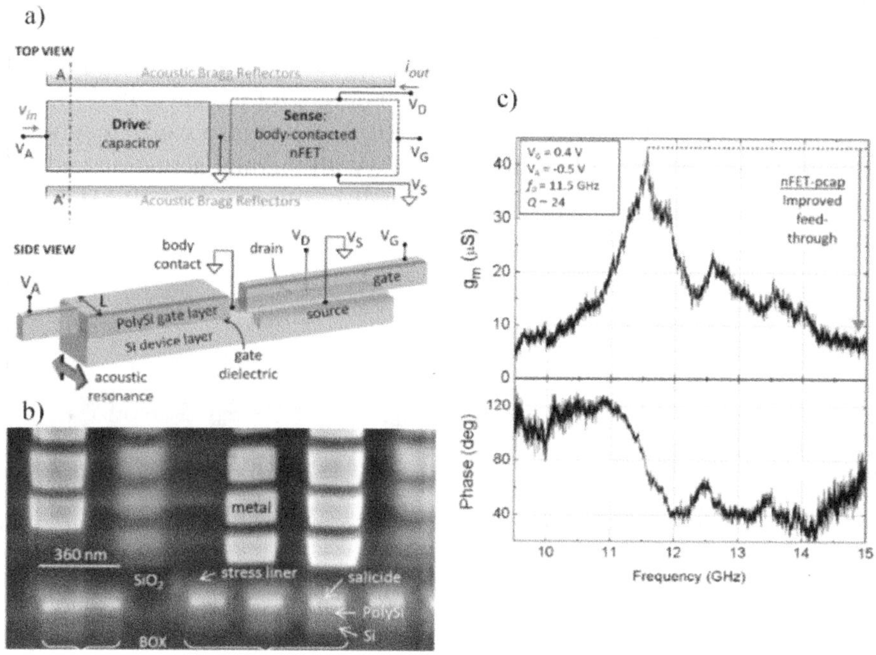

Figure 3.12. *Figures from [MAR 14]: resonant body transistor-type 11 GHz acoustic RF resonators manufactured in the front-end of a CMOS SOI technology of IBM; a) schematic views of the RBT with decoupled actuation and detection; b) FIB image of the CMOS stack and, in particular, the resonant cavities buried under the metallic levels; c) electrical response in terms of transconductance and phase of a resonator based on an actuation capacitor with p-type channel and an nFET transistor. For a color version of the figure, see www.iste.co.uk/duraffourg/nems.zip*

3.3.3. *Alternative approaches*

We will now list other research which does not demonstrate co-integration, but rather describes NEMS which can potentially be co-integrated. For example, [BAR 12c, BAR 14] from the group led by A. Ionescu at the EPFL (Switzerland), concerning nanowire resonators in Si-mono with f_0 in the range of 70–100 MHz, implemented in RBTs

(RB-FET such as [MAR 14, MAR 12], but in this case the resonators are a different type, are not co-integrated, they are released and in a distinct frequency range), see Figure 3.13. [BAR 14], in the same way as [KOU 13], concerns a resonant Si nanowire operating as a junction-less field-effect transistor (FET), using the modulation effect of the depletion charge, induced by the lateral mechanical motion. Results comparing the measurements on the same resonator tend to show that this detection scheme procures a better R_M than PZR detection. This type of resonator can potentially be integrated into a CMOS technology, for instance, by following the process described in [OLL 12] and [ARC 14]. It is also worth mentioning Si-mono resonators, described in [HAL 13], with thermal actuation and PZR detection, which can be implemented in self-oscillation (see [RAH 11]) as candidates for co-integration.

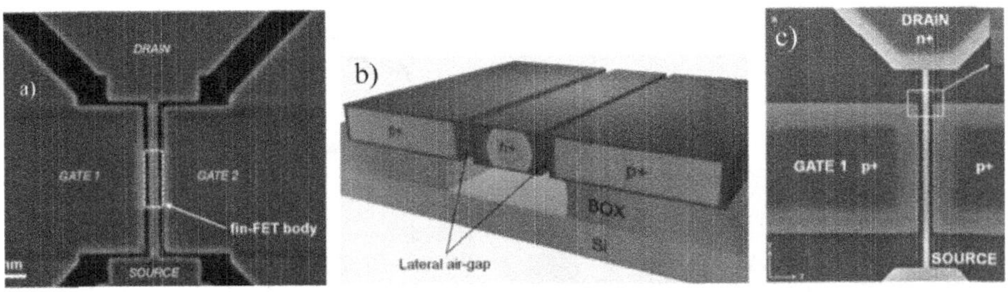

Figure 3.13. *Figures taken from [BAR 12c] and [BAR 14]. Si nanowires implemented in field-effect transistors with a vibrating channel: a) with a junction; c) junction-less. For a color version of the figure, see www.iste.co.uk/duraffourg/nems.zip*

3.4. Conclusions and future perspectives

Co-integrated NEMS-CMOS devices provide not only optimal transduction efficiency by drastically reducing parasitic attenuations and couplings between the resonator and its readout circuit, but also significant gains in size and power consumption of the overall system.

From the perspective of components, co-integration should enable new analog, radio-frequency (RF) and sensor functions to be proposed. In this latter point, it should enable extremely low detection limits to be attained (for instance, < kDa in mass) and it appears to be the best solution for deploying NEMS in arrays with thousands of units or more, and for them to be read individually. Arrays of this type will resemble imagers made up of a set of ultra-sensitive pixels.

From the technological perspective, the trend of NEMS manufacturing processes to converge toward CMOS should become established, which in turn should generate new devices and applications. Among the high potential technological approaches offering an alternative to 3D integration with through silicon vias (TSV), 3D sequential monolithic integration [BAT 12] offers great opportunities, especially in terms of the extreme interconnection density that it could procure. This above-IC technology is based on the molecular bonding of a CMOS wafer with a returned SOI wafer. In this configuration, NEMS in Si-mono can be implemented with doping and crystalline orientation independent of CMOS NEMS.

4

NEMS and Scaling Effects

4.1. Introduction

The move from microelectromechanical systems (MEMS) to nanoelectromechanical systems (NEMS) involves more than the homothetic reduction of the electromechanical properties of the systems. In addition to the scaling laws summarized in Table 1.1, the fundamental effects inherent to nanometric dimensions must be considered. This chapter will briefly describe several examples of scaling effects, some of which will be examined in more detail. These physical laws can be classified into two types: extrinsic and intrinsic phenomena.

Extrinsic phenomena, such as surface interactions, appear outside the material forming the NEMS. In particular, we can cite Casimir/Van der Waals [CAS 48] forces, which are no longer negligible in the presence of electrostatic actuation forces and which are likely to perturb the operation of the NEMS [CHA 01, PAL 05, AND 07]. The Casimir force will be discussed in section 4.2. In the case of NEMS, friction or fluid compression forces occur in the rarefaction regime of the surrounding air. This regime depends on both the intervals separating two structures (fixed or mobile) and the excitation frequencies of the mobile parts. Countless other interactions between interfacing surfaces can occur (for example, read [SIR 10]). All these interactions depend on the conditions of the use of the NEMS (for example, air and vacuum) and the realization modes (the geometry of the interfacing surfaces, their roughness relative to the distance separating them, as well as the materials, can have a strong impact).

Physical laws usually linked to the intrinsic properties of materials, such as electrical and thermal conductions do, however, change at small dimensions. Silicon nanowires like those presented in Figure 4.1 have diameters typically ranging between 5 and 100 nm and have lengths at the micron level. For the smallest diameter, 10 or so silicon atoms can be identified along its diameter. The environment considered here is, therefore, no longer completely continuous with surface conditions that have a strong influence. It is clear that the classic models for thermal and electrical conduction phenomena should by and large be improved. For example, a giant piezoresistive effect was observed [HE 06] in silicon nanowires manufactured in bottom-up technology (see section A1.1). The remainder of this book will focus in particular on this scaling effect that modifies piezoresistive transduction and which is a preferred method for NEMS (see Chapter 2, section 2.2).

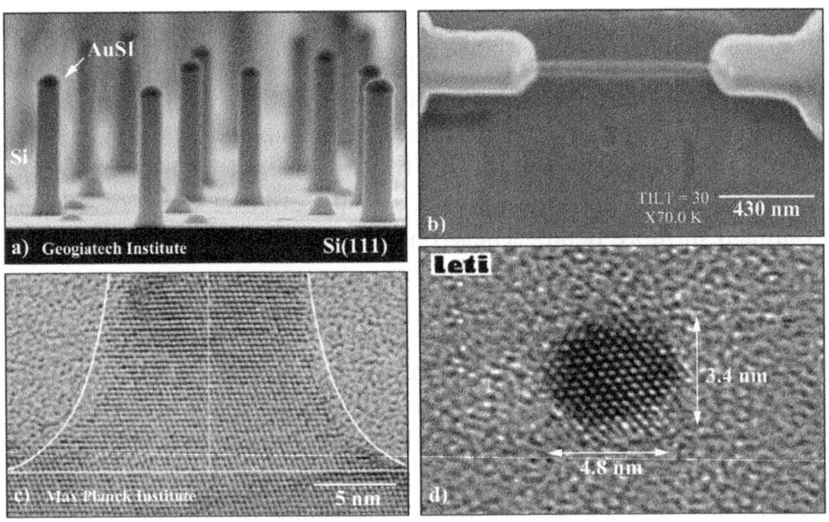

Figure 4.1. *Example of silicon nanowires: a) nanowires built via the vapor–liquid–solid method, Filler research group at GeorgiaTech; b) suspended nanowire built via the nanostructuration of silicon (lithography and etching followed by the successive steps of oxidization/deoxidization); c) TEM cross-section of a vertical VLS nanowire [SCH 09b]; d) TEM cross-section of a horizontal nanowire obtained by etching*

In the same vein, the thermal conductivity of a nanowire decreases with the diameter. At high doping, the nanowire can be seen as an electron gas (or hole gas) associated with a crystalline environment without interaction.

Thermal conductivity is, therefore, the sum of the two electronic $\lambda_{th\_e^-}$[1] and phononic λ_{th_ph} contributions [JU 05]:

$$\kappa = \lambda_{th\_ph} + \lambda_{th\_e^-} \quad \lambda_{th\_ph} = \frac{Cv_g}{3} \frac{1}{l_{ph}^{-1} + w_j^{-1}} \qquad [4.1]$$

$$\lambda_{th_{e^-}} = \frac{1}{3} C_e v_F l_e \qquad [4.2]$$

where C is the thermal capacitance of the crystal, v_g is the group speed of the phonons and l_{ph} is their mean free path in a macroscopic sample. C_e is the thermal capacitance of the electron gas, v_F is the Fermi velocity of the electrons and l_e is their mean free path. The dimension w_j is the width of the nanowire such as to take into account phonon scattering by the surface and l_{ph} is the mean distance traveled by a phonon between two collisions. These collisions occur on the crystal defects (dopant not inserted in the lattice, grain boundary), between the phonons themselves but also with the surfaces of the structure. $C \sim T^3$ at low temperature ($T << \theta_D$, Debye temperature) and $C \sim 3Nk_b$ at high temperature (N is the atom concentration in m^{-3}). The product Cv_g is of the order of $1.7\ 10^9\ W.m^{-2}K^{-1}$ (for example, read [KIT 98]). The mean free path l_{ph} depends on two processes: phonon–phonon (Umklapp process) and phonon-defect interactions. In a solid material, as the displacement of phonons is not restricted by geometric limits, the free path of the phonons only depends on their density and crystalline quality. In a perfect crystal, it is the density of phonons and therefore temperature that sets the free path. For example, in the case of monocrystalline silicon at room temperature, the mean free path of a phonon is about 40 nm.

As a result, a structure requiring small geometric dimensions (i.e. close to or less than the mean free path predicted in the solid material) will modify the mean free path of phonons and consequently the thermal conductivity of the material due to geometric boundaries. The direct consequence for silicon nanowires is a reduction in their thermal conductivity compared with the value expected in a solid crystal (see Figure 4.2). This phenomenon is just as

1 The electrical conductivity of strongly doped nanowires can also be roughly estimated by the Wiedemann–Franz law [KIT 98], $\lambda_{th\_e^-} = \sigma LT$, where L is the Lorenz number ($2.45\ 10^{-8}\ W.\Omega.K^{-2}$). This expression is well suited to noble metals.

significant as the other mechanisms causing phonon collisions, in particular phonon–phonon interactions, which are not much present (see Figure 4.3). The difference in conductivity between a silicon nanowire and a solid material is, therefore, exacerbated in the domain of low temperatures. Conversely, the widths considered here (at room temperature) remain relatively high in relation to the mean free path of electrons or holes. Consequently, there is a sharp decrease in thermal conductivity, which does not damage the electrical conductivity with the size of the wire. As will be seen later, this is no longer true when the nanowire is lightly doped. In this case, electrical conduction is itself affected by the surface states. We will return to this point when discussing the piezoresistive effect in silicon nanowires. In addition, for very small diameters – smaller than 5 nm – quantum confinement phenomena appear to drastically modify valence and conduction bands.

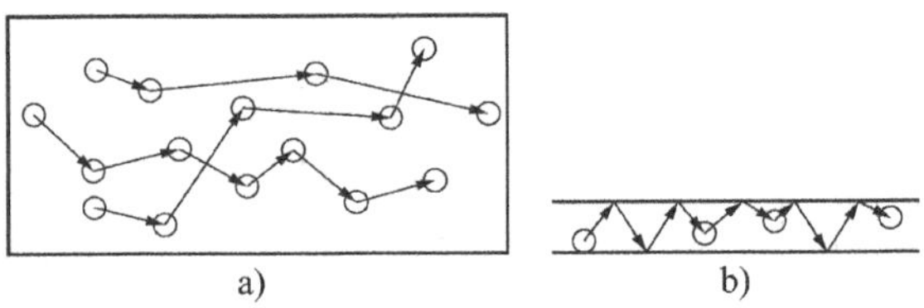

Figure 4.2. *Propagation and collision of phonons; a) in a solid crystal; b) in a nanowire*

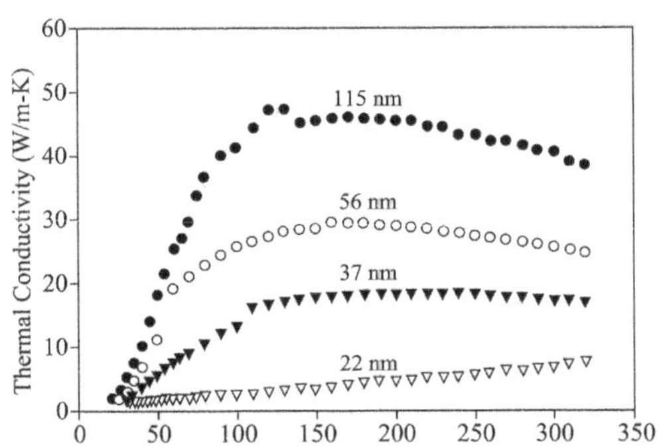

Figure 4.3. *Thermal conductivity of individual nanowires according to temperature for different diameters (graph taken from [LI 03])*

We will now briefly describe the thermoelectrical properties in nanowires. Classically, solid thermoelectrical materials enable a heat flow to be converted into an electrical power and vice versa. For instance, the Seebeck effect corresponds to the appearance of an electrical voltage at the terminals of two junctions formed by two conducting materials of different natures when these materials are at different temperatures. The Peltier effect can be compared with the reciprocal effect. The efficiency of the thermoelectrical material is characterized by its merit factor ZT (adimensional factor), which is a constant parameter characterizing the conversion yield (ratio of the electrical power consumed at the generated heat flow):

$$ZT = S^2 \sigma T / \kappa \qquad [4.3]$$

where S is the characteristic Seebeck coefficient of the material, σ and κ are the electrical and thermal conductivities, respectively. An efficient material will have the highest ZT factor possible.

The expression of ZT shows that the power factor $S^2\sigma$ must be optimized while thermal conductivity κ is minimized (see equations [4.1] and [4.2]). It is, therefore, necessary to try to have a material with high thermoelectrical power that is both a good thermal insulator and a good electrical conductor (see Figure 4.4). Materials are generally alloys based on semiconductors or rare earths and which thus enable parameters $S^2\sigma$, and κ to be optimized (for example, read [SNY 08] and [ROW 05]). Thermoelectrical power S is proportional to the derivative of the state density at the Fermi level. We use the nanostructuration of materials, enabling materials with lower dimensionalities (two dimensional (2D)–0D) [DRE 07] to be obtained. Their state density functions have steeper slopes (Figure 4.5), which therefore improve thermoelectrical power. In reference [HOC 08], the authors demonstrated that the thermoelectrical efficiency of nanowires[2] is exacerbated in relation to the macroscopic junctions in monocrystalline silicon (see Figure 4.6). In this example, the authors use a one-dimensional (1D) rough material (on the scale of the wire diameter) (see Figure 4.6(a)),

2 The thermoelectrical effect observed in nanowires corresponds to the Thomson effect, discovered by William Thomson (and Lord Kelvin), which is often mistakenly confused with the Seebeck effect. It explains the appearance of an electrical current in a material passed by a heat flow while the latter is subjected to a temperature gradient. In contrast, heat is created in a material subjected to a temperature gradient and passed by an electrical current. The Thomson effect connects the two previous effects and can also be applied when there is no junction.

which enabled all parameters to be optimized (S, $S^2\sigma$, and κ). As has been previously explained, the thermal conductivity of the nanowire is thereby reduced in comparison to three-dimensional (3D) silicon (see Figure 4.6(c)). Moreover, roughness enables the scattering of the phonons on the surface to be increased (the surface contributes greatly to the efficiency of thermal resistivity). The thermoelectrical ratio S is, therefore, more favorable to ZT than nanowires without roughness.

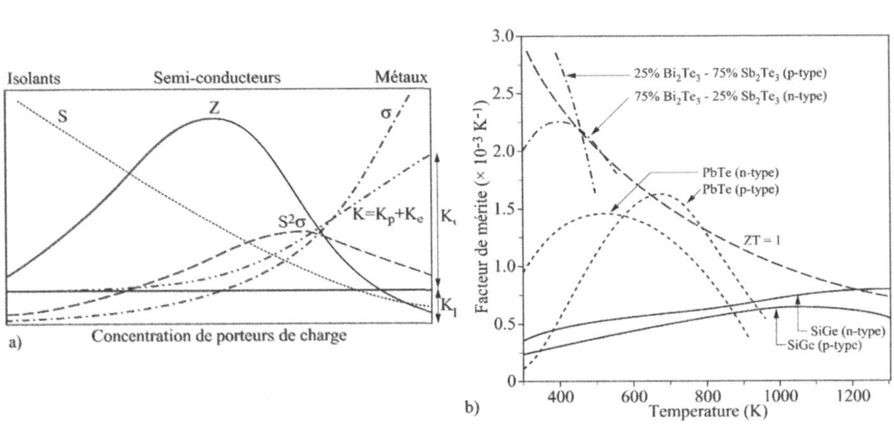

Figure 4.4. *Thermoelectrical materials: a) thermoelectrical power, electrical conductivities and thermal conductivities according to the concentration of carriers: semiconducting materials are more efficient, their thermal conductivity is low at strong electrical conductivity; b) ZT according to temperature for typical thermoelectrical materials: alloys based on semiconductors or rare earths*

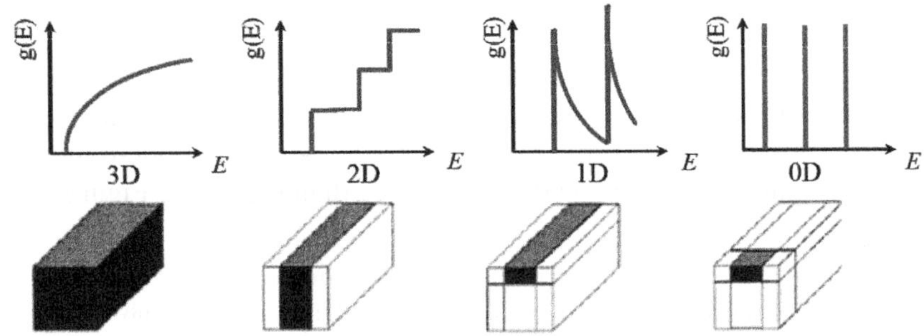

Figure 4.5. *State densities for 3D materials (classic materials), 2D materials (plane structures), 1D materials (nanowires) and 0D materials (quantum boxes) (adapted from [DRE 07])*

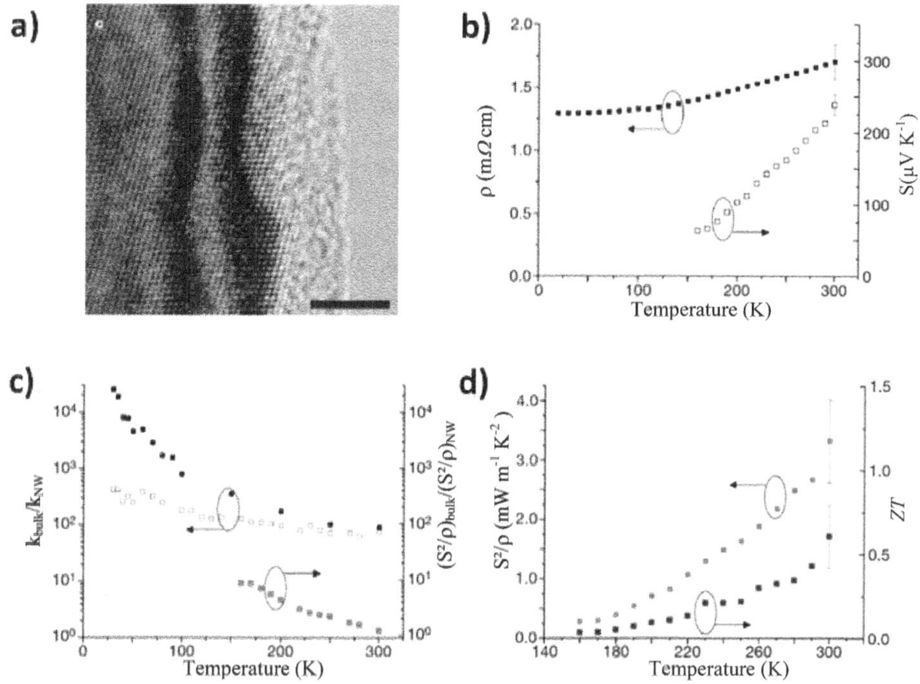

Figure 4.6. *Roughness at the surface of a nanowire and its thermoelectrical performance [HOC 08]: a) TEM cross-section of the surface of a nanowire (scale bar = 4 nm); b) resistivity and thermoelectrical power calculated according to temperature: resistivity increases relatively little in comparison to thermoelectrical power; c) comparison between the 3D solid material and a nanowire: ratio of thermal conductivities in the case of intrinsic solid silicon (full circles) and in the case of strongly doped solid silicon at 1.7 × 10^{19} cm^{-3}(open circles) and ratio $(S^2/\rho)_{bulk}/(S^2/\rho)_{SiNW}$; d) estimation of parameters (S^2/ρ) and ZT according to temperature: ZT~0.6 at room temperature, which is comparable to the values obtained with the complex alloys at high temperatures*

We will conclude this introductory section with a quick study of the mechanical quality factor Q of a nanoresonator according to its dimensions. The quality factor was introduced in equations [2.13] and [2.14] in Chapter 2. In this chapter, fluid damping was cited as the only loss process. There are countless other additional and uncorrelated processes. The global quality factor is written as the inverse of the sum of the inverses of specific quality factors [LIU 05]:

$$Q = \left(\sum_i \frac{1}{Q_i} \right)^{-1} \qquad [4.4]$$

i indicates the specific loss process. Losses can be categorized into two groups:

– *intrinsic losses* inherent to fundamental interactions in the oscillating structure and to imperfections (defects in the lattice surface or volume). Intrinsic losses constitute a sort of loss floor, which it is not possible to go below;

– *extrinsic losses* inherent to interactions between the vibrating structure itself and the surrounding environment (loss at anchoring and hydrodynamic damping).

4.1.1. *Intrinsic losses*

4.1.1.1. *Phonon/phonon interaction*

Phonon/phonon interactions are related to thermoelastic losses when the NEMS is vibrating. If the length of the acoustic wave, set by the vibration frequency of the NEMS, is small in comparison to the mean free path of a phonon, the calculation can be made in the ballistic domain. In this case, acoustic vibration is seen as an acoustic phonon and internal losses can be modeled by elastic or inelastic collisions between "thermal" phonons and acoustic phonons. In the scattering regime, this interaction is defined as a local perturbation of a distribution of phonons (similar to a viscous gas) by the acoustic wave [BRA 85]. In a completely continuous case, the phonons thermalize so quickly in comparison to the acoustic wave that they are assimilated to a local heat flow between the hot zone and the cold zone. This local temperature gradient is created by changes in volumes due to vibration. We are now at the thermoelastic limit, which was studied by Zener [ZEN 38]. In this case, the quality factor is expressed as follows:

$$Q_{th}^{-1} = \frac{\omega\tau}{1+\omega^2\tau^2}\Delta \qquad [4.5]$$

with an amplitude of dissipation:

$$\Delta = \frac{\alpha^2 TE}{C_p}.$$

Here, τ is the relaxation time that corresponds to the time taken by the heat flow to move from the elongated zone to the compressed zone. In the case of a bending beam, τ is equal to $\tau = \frac{e^2 \rho C_p}{\pi^2 \kappa}$ (ρ is the density, C_p is the thermal capacitance and κ is the thermal conductance).

4.1.1.2. *Electron/phonon interactions*

For strongly doped nanoresonators, the movement of the structure causes an oscillation of ions thereby creating an oscillating electrical field. The electrons in the metal can be considered an electron gas that interacts with the electrical field and dissipates energy. These losses originate from the electron/phonon interaction in the structure and should be considered at high frequencies.

4.1.1.3. *Two level systems*

Systems with two levels of energy are actually a simplified mathematical description of trapping phonons by defects (impurities, contaminations and defects in the lattice), which pass the phonon from one level of stable energy to another state of stable energy.

4.1.1.4. *Surface effects*

Surface states for small objects also play a predominant role in the value of the quality factor. These surfaces can be affected by different mechanisms such as the establishment of a water bridge, the creation of dangling chemical bonds or the oxidization of the surface. Surface losses depend on the manufacturing process (thermal treatments, chemical treatments or roughness). Semi-empirical models have, nevertheless, been proposed with varying degrees of success (see [YAN 00] and [WAN 03]).

4.1.2. *Extrinsic losses*

4.1.2.1. *Fluid interactions*

First, the fluid regime in which the beam evolves must be defined. Let us consider a nanowire vibrating according to an out-of-plane eigenmode. Its geometry is described in Figure 4.7. The regime is molecular if the acoustic wavelength of the pressure wave, originating from the bending of the beam λ_g, is considerably lower than the mean free path of the molecule of the ambient gas λ_l. In the extreme opposite case, the NEMS is in a classic viscous fluid environment. When it is between the two domains, the regime is known as the transition regime. We are going to define these two notions and compare their values in a number of cases to provide a point of reference. The acoustic wavelength in the air is simply the ratio of its propagation speed c_g to the frequency considered:

$$c_g = \sqrt{\frac{\gamma R}{M} T}$$

$$\lambda_g = \frac{c_g}{\nu_T} \qquad [4.6]$$

where γ is the ratio of specific heats c_p over c_v, R is the molar constant of the gases and M is the molar mass (for air $c_g = 343$ m.s^{-1}).

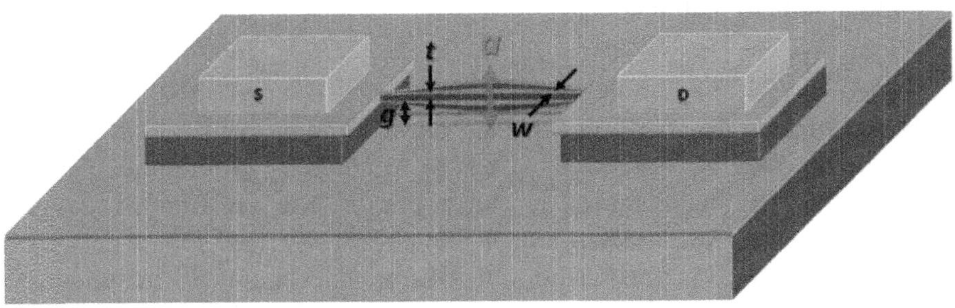

Figure 4.7. *The geometry considered and the notations used: t and w are the thickness and the width of the vibrating beam according to the perpendicular direction at the substrate, g is the distance separating the beam from the substrate, a is the vibration amplitude. For a color version of the figure, see www.iste.co.uk/duraffourg/nems.zip*

It is unusual to define the mean free path of a molecule forming a gas without using a calculation from statistical physics. It is nevertheless possible to define the approximate form by modeling the molecule by a hard sphere with a diameter d_0 (molecular diameter) of [CAL 85]:

$$\lambda_l = \frac{k_B T}{\sqrt{2}\pi d_0^2 P} \qquad [4.7]$$

At ambient pressure (1,013 hPa), the mean free path (of nitrogen) is 90 nm. Figure 4.8 shows the mean free path λ_l of a nitrogen molecule constituting the air depending on pressure compared with the length of the bending acoustic waves λ_T for two beams with different dimensions: L = 5 µm, w = 200 nm, t = 160 nm, f_0 = 56 MHz and L = 0.66 µm, w = 50 nm, t = 50 nm, f_0 = 1 GHz. Roughly, it can be considered that the molecular regime is reached for pressures lower than the pressure corresponding to where curves λ_l and λ_T cross. According to this criterion, the molecular regime is reached for ambient pressures higher than 10 Torrs (1 Torr = 1/760 atmosphere). The molecular regime is reached much faster as the dimensions are small (i.e. resonance frequencies are high). The rarefaction regime of the gas can also be reached in confined spaces when these spaces are small compared with the mean free path λ_l. In this case, the fluid regime is characterized by the Knudsen number *Kn* which is the ratio between λ_l and the separation distance g (the gap between the nanobeam and the fixed part). According to *Kn*, the damping force $F_d = b\dot{y}$ given in equation [2.12] will be calculated using different methods [GAD 99, ZHA 12]. From this definition, five flow regimes can be defined (see Figure 4.9). When *Kn* is lower than 10^{-2}, the flow regime is continuous and can be described by classic laws of dynamics (Euler or Navier Stokes equations). When *Kn* is between 10^{-2} and 1, the gas rarefaction must have previously been taken into account. It is considered, nevertheless, that classic mechanics for calculating the hydrodynamic damping force can still be applied by setting a tangential velocity jump of the gas (known as the "slip flow" regime) as condition to the limit. The condition at the limits requiring the velocity of the fluid to be null on the walls (conventional case) is in fact replaced by the slip boundary condition. This calculation leads to an expression of the force with a variation law g^{-1} instead of a classic law g^{-3}. Siria *et al.* have experimentally shown this change in regime due to the gas rarefaction for small gaps[3] [SIR 10, SIR 09]. Figure 4.10 shows the main lessons learned from these experiments. Primarily dominated by fluid losses, the quality factor follows a g^{-1} law as was predicted by theory. A second approach consists of applying a weighting term to the viscosity μ of the surrounding gas. This term is written according to *Kn* and depends on the interval to which *Kn* belongs [VEI 01].

3 This calculation is made in the quasi-statistical case when the nanostructure is not vibrating. It is applied, by example, to the nano-accelerometer.

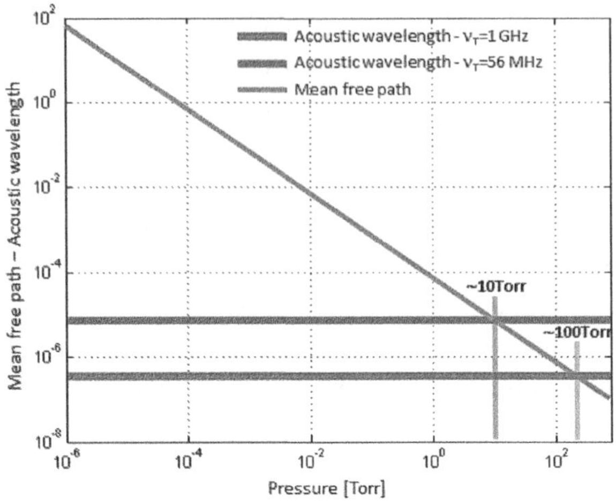

Figure 4.8. *Comparison between the mean free path of a nitrogen molecule (constituent of air) and the acoustic wavelength: for a beam vibrating at 1 GHz, the molecular regime is reached for an ambient pressure of 100 Torr; for a beam vibrating at 56 MHz, the molecular regime is reached for an ambient pressure of 10 Torr. For a color version of the figure, see www.iste.co.uk/duraffourg/nems.zip*

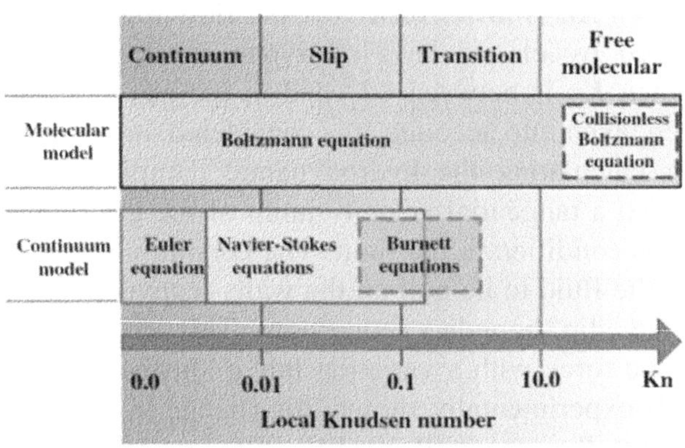

Figure 4.9. *Different flow regimes according to Knudsen number (from [ZHA 12]). For a color version of the figure, see www.iste.co.uk/duraffourg/nems.zip*

When *Kn*>>1, in transition regime or molecular regime, only molecular models are used to resolve a problem at N body. A statistical approach describes how to resolve this complex problem at N coupled equations. The trajectory of each gas molecule is not considered directly, but rather the probability density of finding a molecule at instant *t*, at position (x, v).

Readers might find it useful to refer to works about statistical physics [ZHA 12] to gain a deeper understanding of the topic. After a number of non-trivial mathematical manipulations (which fall beyond the scope of this work), the general form to be integrated is the Boltzmann equation. In the case of small volumes (typically in microfluidics) and for very low integration durations (~1 ns), molecular dynamic methods can determine in a deterministic way the position and the speed of each molecule by considering, where necessary, the laws of quantum mechanics[4]. In extremely simple cases (like the geometry described in Figure 4.9(b) representing a mobile rigid plane in front of a support), the quality factor inherent to fluid damping can be analytically determined [BAO 02].

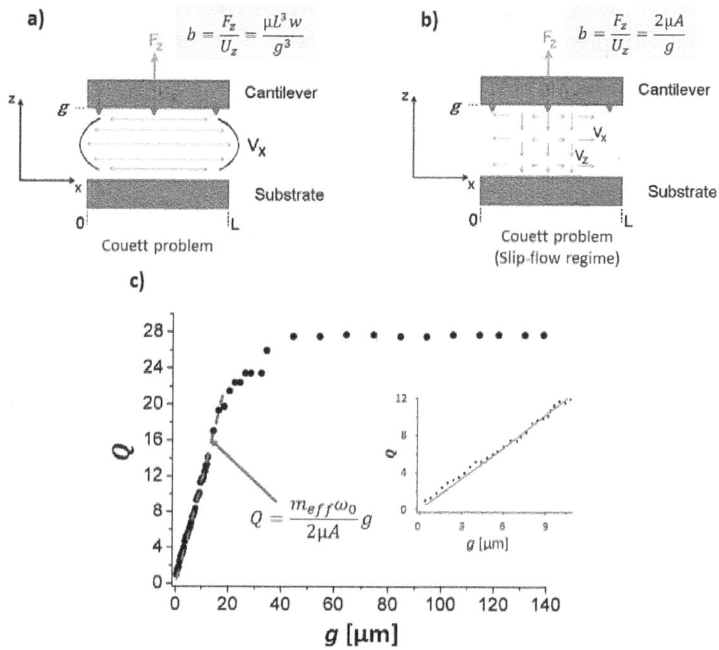

Figure 4.10. *a) Confinement scheme of the fluid in continuous regime and hydrodynamic force in d^{-3}; b) confinement scheme of the fluid in "slip flow" regime and hydrodynamic force in d^{-1}; c) quality factor of a nanomembrane measured by an interferential method, the light provided by the fiber optics facing the NEMS; the gap d between the fiber and the NEMS is variable (figure adapted from [SIR 09]). For a color version of the figure, see www.iste.co.uk/duraffourg/nems.zip*

4 Hamiltonian interactions (molecules/molecules and molecules/solid limits) like the Lennard–Jones potential, for example, and Hamiltonians corresponding to the internal energy of the molecules (relative rotation and relative vibration of the atoms).

4.1.2.2. *Dissipation of mechanical energy in the support*

When this nanobeam vibrates according to one of these bending eigenmodes (see section 2.1), some of the elastic energy is transmitted to the support. In other words, there is a coupling between the vibration modes of the beam and the eigenmodes of the elastic waves capable of propagating in the support. In general, the coupling rate cannot be calculated analytically. The coupled equations, corresponding to the bending of the beam and the propagation of acoustic waves in the supports in three dimensions, must be resolved simultaneously. It is, therefore, necessary to use dynamic digital simulations based on finite element method (FEM). When the beam is fine (thickness << than the wavelength of transverse acoustic waves propagating in the support), the calculation can be made in two dimensions in the plane of the beam. Supposing that coupling remains low because the thickness of the beam is low, a perturbative approach can be used to resolve the equations. Like in the case of losses due to surface defects, semi-empirical models have been introduced with varying degrees of success (for example, read [YAN 00]).

The following sections will provide details about two physical phenomena symbolic of the size reduction effect: (1) the Casimir force, which is extrinsic to NEMS and chiefly depends on the gap g (see Figure 4.7) and (2) the modification of electrical conduction and its associated piezoresistive effect, which are intrinsic to NEMS. We will then focus on optomechanical systems. Quantum effect and back-action, which we briefly discussed in Chapter 1, will be discussed in this case.

4.2. Near field effect in a nanostructure: Casimir force

4.2.1. *Intuitive explanation of the Casimir force*

The Casimir force is one of the physical manifestations that can be observed in the macroscopic world of the effect of quantum fluctuations of the vacuum. This force was calculated in 1948 by Casimir [CAS 48] and accurately measured for the first time by the team led by Lamoreaux [LAM 97] in 1997. It has since been the object of a vast number of theoretical and experimental studies and as many publications [LAM 07b, KLI 09]. The Casimir force F_c per surface unit between two uncharged semi-infinite without roughness plates that are perfectly parallel and have perfect electrical conductivity, separated by a distance d, is expressed as follows:

$$F_c = \frac{\pi^2 \hbar c}{240 d^4} \qquad [4.8]$$

where c is the speed of light and $\hbar$ is the Planck constant, sign of the quantum origin of this force.

The Casimir force is an attractive force present between the two interfacing planes. It originates from the quantum fluctuations of the vacuum. Vacuum in quantum mechanics corresponds to the physical state of the fields, regardless of their nature, at their minimum energy state. The electromagnetic field decomposes according to its eigenmodes which all resemble harmonic oscillators of energy U_l so that its total energy U is written (in second quantification of the field, for example, read [GRY 10]):

$$U = \sum_l U_l = \sum_l \left(n_l + \frac{1}{2} \right) \hbar \omega_l \qquad [4.9]$$

where n_l is the quantum number of the mode l also known as a photon at pulsation ω_l of the wave. In this approach, the electromagnetic quantum vacuum corresponds to a zero photon field state $n_l = 0$, with a minimum energy $U = \sum_l \hbar \omega_l / 2$.

To improve understanding, any monochromatic field E_l can be rewritten according to two components in cosine and sine quadrature:

$$E_l = E_1 \cos(\omega t) + E_2 \sin(\omega t) \qquad [4.10]$$

These components in quadrature have a mean value of null but the standard deviations ΔE_l and ΔE_2 are not null, thus verifying the following uncertainty[5]:

$$\Delta E_1 \Delta E_2 \geq E_0^2 \qquad [4.11]$$

5 This uncertainty is similar to that defined by Heisenberg.

where E_0^2 is the minimum energy equal to $\hbar\omega_l/2$. In addition, equation [4.11] shows that the components in quadrature cannot be perfectly defined simultaneously. This minimum energy E_0^2 for a mode l corresponds to the energy of the vacuum fluctuations (without photon) in a space without boundaries.

A first intuitive approach explains the Casimir force between the two planes as follows. The two planes constitute a Fabry–Pérot-type resonant cavity (see Figure 4.11(a)). This optical cavity acts as a spectral filter in which only the eigenmodes ω_c of this cavity can subsist. This spectral filtering is described by the transfer function f shown in Figure 4.12(b). The eigenmodes of the cavity ω_c are proportional to $c/2d$ (c is the speed of light in free propagation). The optical cavity acts as a spectral filter only preserving the eigenmodes ω_c, the pressure exerted by the quantum fluctuations of the vacuum outside the cavity P_{out} is larger than the pressure due to the intra-cavity fluctuations P_{cav} (see Figure 4.11(b)). The two planes, therefore, tend to move closer to each other. This pressure difference is at the origin of the Casimir force.

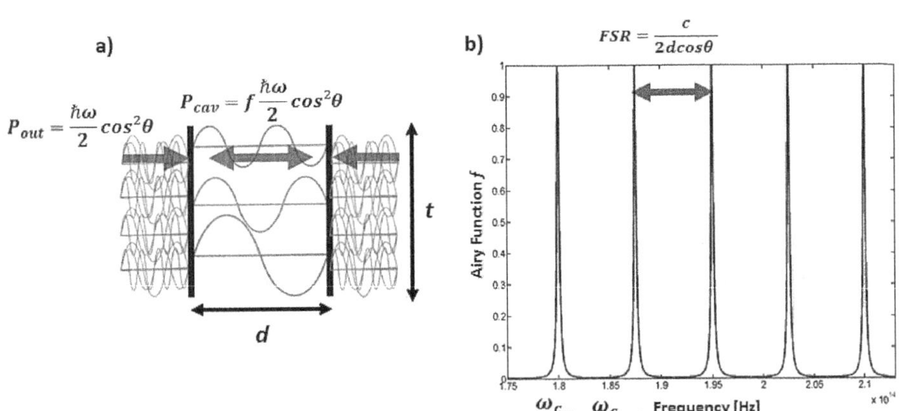

Figure 4.11. *a) Scheme of the Fabry–Pérot optical cavity. P_{out}: external pressure exerted by the set of the free modes; P_{cav}: internal pressure exerted by the eigenmodes of the cavities; the optical cavity acts as a filter and only preserves the eigenmodes whose angular frequency is proportional to c/2d, the internal pressure is lower than the external pressure exerted by the set of fluctuating modes; filtering is taken into account by the Airy function f which represents the energy transfer function of the optical cavity; b) Airy function f; FSR: free spectral range separating two consecutive optical eigenmodes. For a color version of the figure, see www.iste.co.uk/duraffourg/nems.zip*

4.2.2. *The problem*

The following question has sparked a great deal of debate: is the Casimir force a limiting factor that is really visible in nanosystems? The question is worth asking in relation to structures as small as those presented here. Given that the characteristic dimensions of NEMS structures are often still significant compared with the atomic scale, the Casimir force should, strictly speaking, be taken into consideration in mechanical equations when a mechanical structure and a fixed part (see Figure 4.7) are separated by a distance less than 200 nm. Nevertheless, the formalism for calculating Casimir force must be adapted from a theoretical context for thick metallic mirrors to thin layers of silicon doped at varying levels. Precursory research considering the perfect expression of Casimir force has been published in the literature (for example, read [BAT 07, DIN 01] and [GUO 04]). We will now introduce a formulation of the force adapted to silicon NEMS which enables its impact on dynamic operation to be better predicted. Several examples of measuring this force on accelerometers will then be presented.

The structure is schematized in Figure 4.12. The accelerometer is composed of a freely moving proof mass fixed to the substrate by four bending arms acting as four 1D springs. As a result of acceleration, the electrostatic combs joined to the mass are displaced with respect to the interfacing fixed combs. Acceleration is, therefore, measured through the capacitance variation between the electrostatic combs. The physical and morphological characteristics are summarized in Table 4.1.

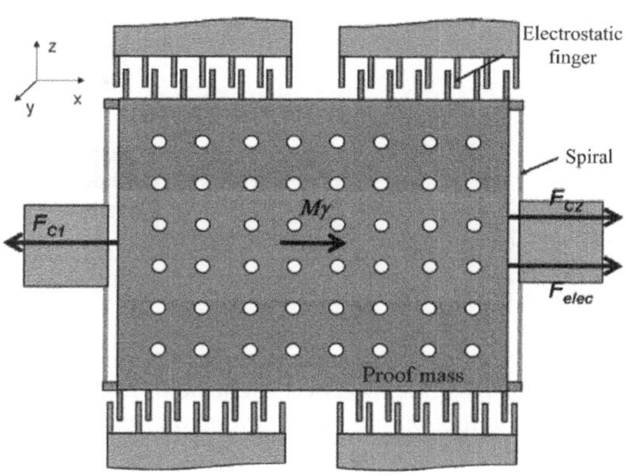

Figure 4.12. *Scheme of an accelerometer (the proof mass has holes so that the sacrificial layer below can be removed and thus suspended) and definition of the forces present – Casimir forces F_{C1} and F_{C2} and electrostatic force F_{elec}*

Overall structure	Thickness (nm)	160
Proof mass	Width (µm)	75
	Length (µm)	150
Bending beam	Width (nm)	50
	Length (µm)	5
	Number	4–6
Electrostatic combs	Number	4
Electrostatic fingers	Width w (nm)	100
	Length l (µm)	5
	Gap d (nm)	100
	Number	125
	Doping (cm^{-3})	10^{20}

Table 4.1. *Typical characteristics of a nano-accelerometer (see Figure 4.13)*

The components are manufactured on a silicon on insulator (SOI) wafer. The structuration was realized by hybrid e-beam/deep UV (DUV) lithography and plasma etching. Figures 4.13(a) and (b) show scanning electron microscopy (SEM) photographs of the final result. The electrostatic parts have been doped around 10^{20} cm^{-3}. For more details on the manufacturing process, please refer to [AND 06].

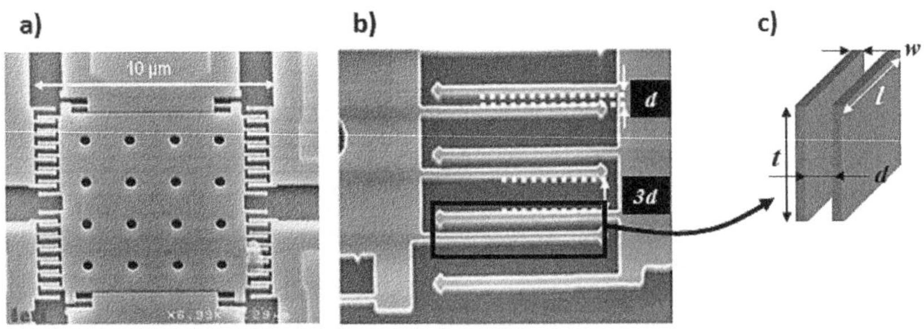

Figure 4.13. *a) SEM photograph taken from above of a typical accelerometer; b) zoom on the electrostatic fingers; c) scheme of two interfacing teeth considered as two optical mirrors constituting an optical resonant cavity*

The electrostatic combs behave like a network of interfacing mirrors separated by a mean distance d of 100 nm. They are unique as they have a small thickness w and a structuring material of doped silicon. There is a Casimir force between each pair of mirrors, which tends to bring them closer together. The resultant of this force acts on the mobile part (the proof mass). Its value is, therefore, compared to the values of the other forces (mechanical and electrostatic).

4.2.3. *Rigorous calculation of the Casimir force between two silicon slabs*

The abrupt calculation of the Casimir force in NEMS with this simple (and elegant) expression of Casimir force gives a value that is very different to reality. Indeed, in the majority of cases, and in particular in nano-accelerometers, plates have a low thickness. Incidentally, this thickness has to be compared with the skin depth, but we will return to this point later. Silicon is also far from being a perfect conductor; this is an essential parameter to include in calculations. With regard to roughness, its root mean square (RMS) value must be compared with the mean value of the gap d [NET 05a, NET 05b]. As is shown in Figure 4.14, roughness of 1 nm remains very low with regard to a gap of 100 nm and leads to a negligible correction of the so-called ideal force given in [4.8], in comparison to other contributions. The effect of temperature is not included in the study as it remains low for distances significantly lower (mean gap of 100 nm) than the thermal wavelength evaluated at 7 μm at room temperature [BOS 00, KLI 09].

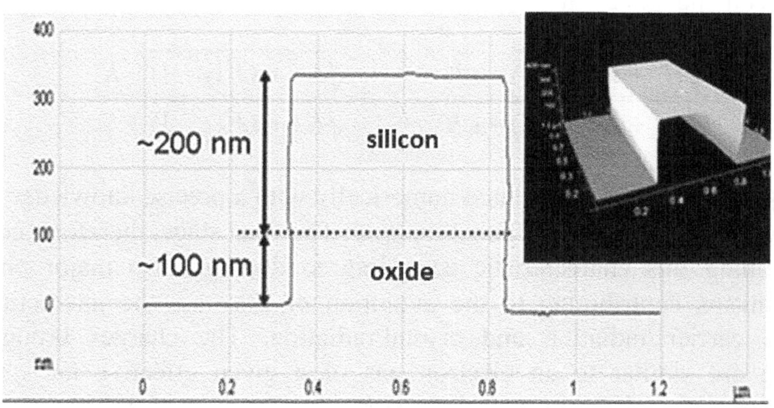

Figure 4.14. *Morphology of a silicon finger on oxide (3D AFM) – atomic force microscopy (AFM) image of an electrostatic finger (picture from LETI). For a color version of the figure, see www.iste.co.uk/duraffourg/nems.zip*

To take derivation from an "ideal" Casimir force into account, the correction coefficient η, which is the ratio between the real pressure taking into account the characteristics of the slabs (doped silicon and thickness w of the slabs) and the pressure given in equation [4.8], is usually defined:

$$\eta = \frac{\langle \hat{P} \rangle}{F_c} \qquad [4.12]$$

P can be interpreted as the resultant of the pressures of quantum fluctuations of the vacuum applied to the optical cavity formed by the two slabs (internal and external pressure to the cavity). S. Reynaud developed a very elegant method of calculating P based on an integral defined in the complex plane for ease of calculation:

$$\eta = \frac{120}{\pi^4} \sum_m \int_0^{+\infty} dKK^3 \int_0^K \frac{d\Omega}{K} f(r_m^2, K, x)$$
$$f(r_m^2, K, x) = \frac{r_m^2 (1 - e^{-2x})^2 e^{-2K}}{(1 - r_m^2 e^{-2x})^2 - r_m^2 (1 - e^{-2x})^2 e^{-2K}} \qquad [4.13]$$

where K is the optical phase shift produced by a return or round trip projected on the propagation axis z and normalized. Ω is the normalized pulsation. The expression of x corresponds to the phase term in each slab normalized at d. $r_{s,p}$ are the reflection coefficients in intensity at the air/silicon and silicon/air diopter for polarizations s and p, respectively:

$$r_p = \frac{\sqrt{K^2 + \Omega^2(\varepsilon - 1)} - \varepsilon K}{\sqrt{K^2 + \Omega^2(\varepsilon - 1)} + \varepsilon K} \qquad r_s = \frac{\sqrt{K^2 + \Omega^2(\varepsilon - 1)} - K}{\sqrt{K^2 + \Omega^2(\varepsilon - 1)} + K} \qquad [4.14]$$

Equation [4.13] is calculated numerically with a precise knowledge of the geometry and the dielectric function $\varepsilon(\omega)$. The final stage, therefore, consists of defining this characteristic according to doping. Two major physical mechanisms contributing to the evolution of $\varepsilon(\omega)$ are the interactions of charge carriers/radiation and crystal/radiation. The charges brought by doping are similar to an electron gas of a given effective mass whose contribution is relatively well defined by the Drude model. The crystalline contribution is calculated from tabulated data (refractive indices and absorption coefficient) [PAL 85]:

$$\begin{aligned}
&\varepsilon(\omega) = \varepsilon_{\text{int}\,rinsic}(\omega) - \frac{\omega_p^{\;2}}{\omega(\omega - i\gamma)} \\
&\omega_p = \sqrt{\frac{Ne^2}{\varepsilon_0 m^*}} = \frac{2\pi c}{\lambda_p} \\
&\gamma = \frac{Ne^2\rho}{m^*} \\
&\varepsilon_{\text{int}\,rinsic}(\omega) = \varepsilon_1(\omega) + i\varepsilon_2(\omega)
\end{aligned} \qquad [4.15]$$

where ρ is the resistivity of the silicon, N corresponds to the level of doping considered. ω_p (plasma frequency) and γ (the relaxation coefficient) are calculated by considering the actual conduction mass of the electron $m^* = 0.34m_e$ for a doping P (m_e is the mass of the electron) [DUE 06]. Table 4.2 shows some plasma frequencies for different concentrations of boron atoms (doping P).

Doping	ω_p (x10^{12} rad.s^{-1})	γ(x10^{13} s^{-1})	ρ (Ω.cm)
1.1.10^{15} cm^{-3}	3.20	1.182	13
1.3.10^{18} cm^{-3}	110.18	3.760	3.5.10^{-2}
1.4.10^{19} cm^{-3}	361.59	7.688	6.8.10^{-3}
10^{20} cm^{-3}	966.38	9.918	1.2.10^{-3}

Table 4.2. *Plasma frequency and relaxation coefficient for different levels of doping*

Finally, equation [4.15] gives a good representation of the dielectric function provided that the semiconductor is not degenerated. All that remains to be done is to define the permittivity in the complex plane by applying the Kramers–Kronig relation as follows:

$$\varepsilon(i\xi) = 1 + \frac{2}{\pi}\int_0^{+\infty} \frac{x\varepsilon_2(x)}{x^2 + \xi^2}dx \qquad [4.16]$$

We can test that the dielectric function of the intrinsic silicon defined in the complex plane [4.16] tends toward 11.7 when frequency tends toward 0, and that it tends toward 1 when frequency tends toward infinity (see Figure 4.15). To take doping into account, [4.16] then needs to be added to the imaginary part of the Drude function [4.15][6]. A similar calculation made on gold enables us to state that the dielectric function of silicon doped at 10^{20} cm^{-3} has a similar behavior to that of gold. This material is used as a reference as it has been intensely studied and enables us to validate the results obtained on Si.

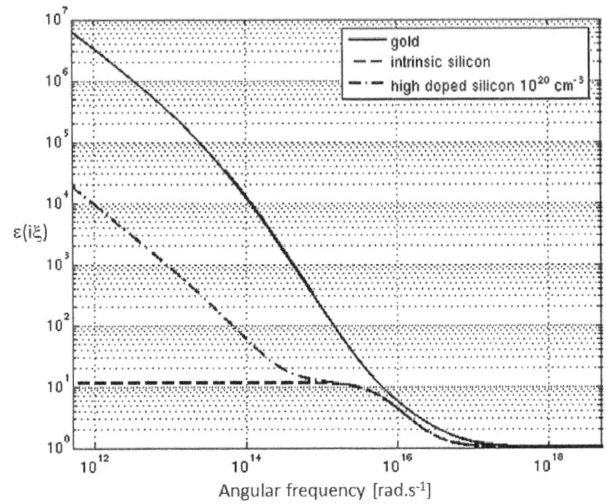

Figure 4.15. *Global permittivity according to the angular frequency expressed in the complex plane for intrinsic silicon and strongly doped silicon (p-type) (see equations [4.15] and [4.16]) – comparison with the permittivity of gold calculated from tabulated data [PAL 85]*

It is important to consider a wide interval ranging from low to high frequencies so that all the modes contributing toward the Casimir force can be taken into account. In particular, for silicon, it seems that frequencies lower than 10^{12} rad.s^{-1} no longer contribute toward the force as the energies in play are negligible. For frequencies higher than 10^{18} rad.s^{-1}, the slabs become transparent and their contribution to the Casimir force also tends toward 0. Finally, we need to plot the correction factor at the ideal force according to the distance separating two interfacing slabs by considering the typical values met in NEMS ranging between several nanometers and 1 μm (see Figures 4.16(a) and (b)).

6 It is also possible to add up the intrinsic permittivity and the permittivity from the Drude model and to realize the Kramers–Kronig transform on the imaginary part of the set. Both methods are equivalent.

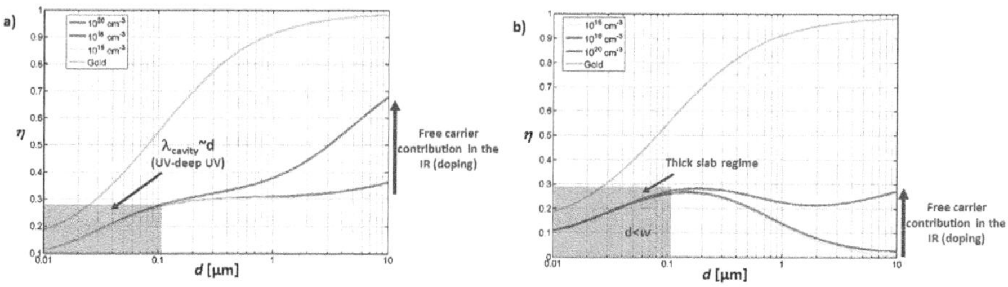

Figure 4.16. *a) Corrective factor according to distance for thick plates (10 μm) compared to the gap value – gold is used as a reference in relation to different levels of silicon doping; b) corrective factor according to distance for slim slabs (100 nm) compared to the gap value – three cases were studied: moderately doped silicon, strongly doped silicon and gold, used as a reference material. For a color version of the figure, see www.iste.co.uk/duraffourg/nems.zip*

Let us look at thick slabs (Figure 4.16(a)). For wide gaps, the corrective factor of the strongly doped silicon tends to increase according to the separation distance. Intrinsic silicon tends toward an asymptotic value of 0.3. For small distances below 100 nm, these two curves move closer together. The level of doping has a strong impact on the Casimir force for large distances. The frequencies selected by the resonant cavity are in essence proportional to $1/d$. At a small separation distance, only the high frequencies (for instance, deep UV for 10 nm) are selected, whereas infrared (IR) frequencies are present for large distances (for instance, IR for 200 nm). The free charge carriers brought by doping contribute in the IR (through $\varepsilon(i\xi)$). It is, therefore, natural to state that the effect of doping is only visible for wide gaps that put the eigen wavelengths in the IR into play. If you need convincing of this, just remember the plasma frequency or look at Figure 4.15 at low frequencies.

We will now concentrate on the corrective factor η according to separation distance for slim slabs (see Figure 4.16(b)). In addition to doping, the width of the slabs has a strong impact on the value of η. This phenomenon can easily be explained by comparing the width of the slab (100 nm) to the skin depth δ_a at the frequencies considered [DUR 06, LAM 07]:

$$\delta_a = \frac{c}{\omega k_a} \tag{4.17}$$

where ω is the angular frequency, c is the speed of light and k_a is the absorption coefficient of the material from the tabulated data in [PAL 85]. δ_a is plotted (see Figure 4.17) for the three cases studied (intrinsic silicon, strongly doped silicon and gold). The thickness of the slabs is also indicated to illustrate this. The reflection coefficient of a silicon slab is non-negligible in the narrow spectral band [5.10^{15}–5.10^{16} rad.s^{-1}]. The resonant cavity is only selective on this band. The Casimir force will, therefore, decrease compared to the case of the thick slab. This effect is all the more clear when the slabs are far apart because the frequencies considered will be increasingly low. In contrast, a 100 nm gold slab will reflect on practically any range of frequencies and the Casimir force will remain very similar to that of thick slabs[7].

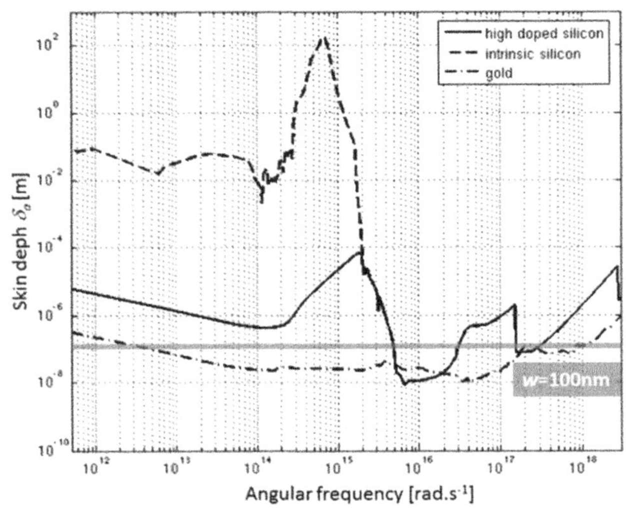

Figure 4.17. *Thickness of skin δ according to angular frequency – the blue line represents the width of the silicon slab typically found in NEMS compared to δ_a (figure adapted from [DUR 06])*

These early results [DUR 06] were completed by the research group led by Lambrecht [LAM 07a, JOU 09]. The effects of geometry and doping are summarized by the 3D curves presented in Figure 4.18.

7 Low frequencies, which are "rejected", bring little energy and only induce a minor correction of the force according to thickness. At very high frequencies, gold will become transparent regardless of the thickness, which will induce no major difference between a slim and a thick slab.

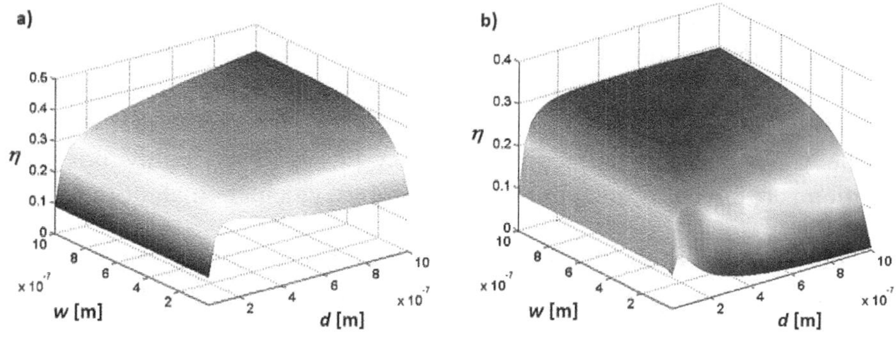

Figure 4.18. *Corrective factor η according to the gap d and the width of the slabs w: a) strongly doped silicon (10^{20} cm^{-3}); b) intrinsic silicon. For a color version of the figure, see www.iste.co.uk/duraffourg/nems.zip*

4.2.4. *Impact of the Casimir force in a nano-accelerometer*

At this stage of the discussion, we need to take a "mechanical" interlude and equate the balance of the forces applied on the proof mass [HEN 07]. This calculation is made for illustration purposes and as such we will outrageously simplify the problem by considering that the set of the proof mass and electrostatic combs is rigid and that the suspensions are without mass. According to Newton, we therefore have:

$$M\ddot{x} = M\gamma + F_{stiffness} + F_{elec} + F_{Casimir} \quad [4.18]$$

where x is the displacement of the center of gravity of the proof mass, γ is the acceleration to be measured and $F_{stiffness}$, F_{elec} and $F_{Casimir}$ are the reaction force of mechanical suspensions, the electrostatic force used to read the capacitance variation and the Casimir force calculated for the set of combs (see equations [4.12] and [4.13]), respectively. By expressing all these forces according to geometric, material and electric parameters, equation [4.18] can be rewritten as follows:

$$-M\gamma + N_p\left(\left(12.40\frac{EI}{L^3} + 1.22\frac{\sigma S}{L}\right)x + 0.74\frac{ES}{L^3}x^3\right) = \\ -\frac{2NS\hbar c\pi^2}{240}\left[\frac{\eta(e_0+x)}{(e_0+x)^4} + \frac{\eta(3e_0+x)}{(3e_0+x)^4}\right] + \frac{2NS\hbar c\pi^2}{240}\left[\frac{\eta(e_0-x)}{(e_0-x)^4} + \frac{\eta(3e_0-x)}{(3e_0-x)^4}\right] \quad [4.19]$$

where N_p is the number of suspensions considered. E, I, S and σ are the silicon's Young modulus, the beam's moment of inertia, its section and internal constraint, respectively. ε_0 is the permittivity of the vacuum. x is the position of the center of gravity of the proof mass. η is the corrective factor on the Casimir force defined in equation [4.19]. This highly nonlinear equation is resolved by an appropriate and robust calculation algorithm.

Figure 4.19 gives the equilibrium position of the center of gravity of the proof mass according to acceleration for different scenarios. It should be noted that the abrupt application of the Casimir force (without corrective factor) would lead us to conclude that no stable position exists. This conclusion has never been experimentally verified. The dynamic range of the theoretical accelerometer[8] spans from 0–50 g to 0–30 g by introducing the real Casimir force (i.e. by including the weighting factor) in the mechanical equations. By also including the nonlinear effects, which tend to counterbalance the Casimir effort, the maximum accessible acceleration shifts from 30 to 37 g, corresponding to a 25% correction compared to the 50 g initially predicted.

A second parameter to be studied is pull-in voltage which causes the electrostatic combs to bond with each other. For this to happen, electrostatic force must be introduced into the previous equation, which is induced by the readout voltage, enabling the capacitance variation to be evaluated:

$$N_{beam}\left(\left(12.40\frac{EI}{L^3}+1.22\frac{\sigma S}{L}\right)x+0.74\frac{ES}{L^3}x^3\right)=$$

$$N\varepsilon_0 SV^2\left[\frac{1}{(e_0-x)^2}-\frac{1}{(3e_0+x)^2}\right] \qquad [4.20]$$

$$-\frac{2NS\hbar c\pi^2}{240}\left[\frac{\eta(e_0+x)}{(e_0+x)^4}+\frac{\eta(3e_0+x)}{(3e_0+x)^4}\right]+\frac{2NS\hbar c\pi^2}{240}\left[\frac{\eta(e_0-x)}{(e_0-x)^4}+\frac{\eta(3e_0-x)}{(3e_0-x)^4}\right]$$

where V is the readout voltage applied to electrostatic combs.

Figure 4.20 shows the equilibrium position according to this readout voltage in the same conditions as before. Once again, the Casimir force produces a softening mechanical effect which balances out the hardening effect of mechanical nonlinearities. A 10% variation in the pull-in voltage is

8 i.e. the maximum acceleration that can be reached before bonding.

visible. It appears that anyone conducting the experiment finds it difficult to observe this effect without taking many preliminary precautions.

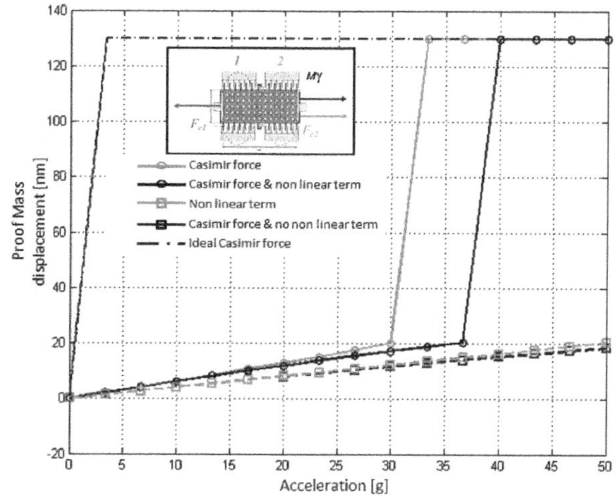

Figure 4.19. *Equilibrium position of the proof mass according to acceleration. For a color version of the figure, see www.iste.co.uk/duraffourg/nems.zip*

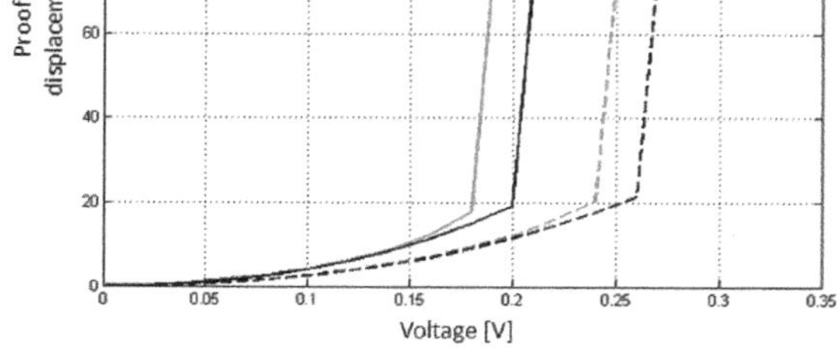

Figure 4.20. *Equilibrium position according to readout voltage. For a color version of the figure, see www.iste.co.uk/duraffourg/nems.zip*

In addition to the Casimir force measurements conducted on an atomic force microscopy (AFM) instrument [KLI 09], the impact of the Casimir force on the operation of an accelerometer has never been formally observed. A sample (the morphology is provided in Figure 4.12) is placed in a room under nitrogen flow, acting as a Faraday cage. The nitrogen prevents the oxidization and contamination of the electrostatic combs. The capacitance variation according to an actuation voltage applied to a pair of electrostatic combs is measured using an impedance meter HP4284. Electrostatic capacitance is read by an alternative current (AC) readout voltage set at 1 V (peak-peak) to 1 MHZ applied to the second pair of combs. Parasitic impedances are eliminated by calibration by first measuring in open circuit and then in short-circuit. The decision is made for readout frequency to be higher than the mechanical cutoff frequency of the nano-accelerometer so that its dynamic response is almost null (we also seek to minimize 1/f-noise). Integration time is set at 100 ms. The results of the measurements in these conditions are presented in Figure 4.21. The nominal capacitance measured was 13.263 aF, whereas the theoretical capacitance was 13.615 aF. The capacitance variation is 40 aF for an actuation voltage varying from 0 to 0.2 V. The noise of the measurement chain (including the NEMS) is evaluated at 1 aF. These experimental data are compared with the theoretical calculations using the previous electromechanical model (for a doping of $\sim 10^{20}$ cm^{-3}). Theoretical curves that include a stress of 5 MPa appear to match the measurements. Having said that, internal mechanical constraints appear to be the key approximation parameter for real predictability. Finally, measurements at low actuation voltages remain very dispersed.

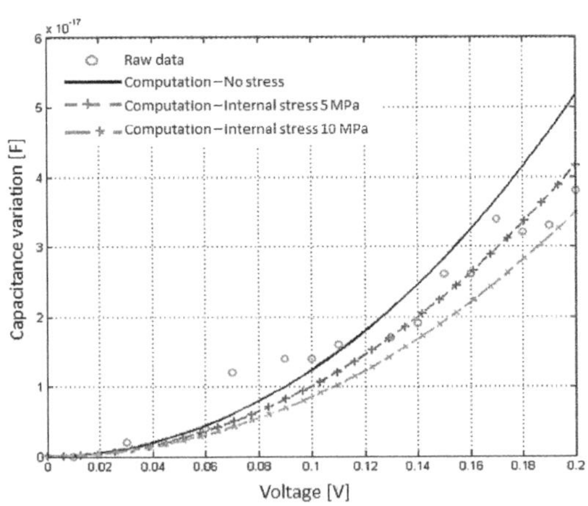

Figure 4.21. *Capacitance variation according to readout voltage. For a color version of the figure, see www.iste.co.uk/duraffourg/nems.zip*

4.2.5. *Conclusion*

In the previous example, the theoretical impact of the Casimir force seems clear. Experimentally, the demonstration of its influence has not, however, been totally conclusive. The main reason for this is the value of the capacitances, which are at the limit of the measurement threshold of the devices. The effect is, therefore, in part due to measurement error. More recently, other research [ARD 12] has more clearly demonstrated the effect of this force in accelerometers by measuring capacitance variations. The impact of the Casimir force has been more widely studied on membranes, plates and switches [GUO 04, BAT 07, PAL 05].

Since then, intense research activity has been undertaken to study how to control Casimir force using the other physical parameters of the materials by modifying the value of the Casimir force or its sign, such as the effect of the roughness or the nanostructuration of the surfaces [ZHA 09, MUN 09, INT 13]. It is understood that this nanostructuration, in a Bragg grating, for example, diffracts the fluctuation fields of the vacuum according to the diffraction orders that are function of the pitch of the grating. Only the wave vectors K of equation [4.13] corresponding to the diffraction orders have a chance to participate toward Casimir force. According to the distance between the planes, the force can change nature and shift from attractive force (linking) to repulsive force (anti-linking) or vice versa. At its extreme, the system could behave as an oscillator. In recent years, Casimir force has also been studied in the context of more exotic meta-material-type optical materials, enabling us to see a new levitation principle by Casimir force on the horizon [PAP 14, LAH 13].

4.3. Example of "intrinsic" scaling effects: electrical conduction laws

We will first focus on electrical conductivity and then on giant piezoresistivity observed with silicon nanowires. We will then conclude with a comparison between piezoresistive transduction and transduction by junction-less transistor (see Chapter 3, section 3.3.3).

4.3.1. *Electrical resistivity*

Let us consider once again the nanowire shown in Figure 4.1. Different processes contribute to the change in electrical resistivity.

4.3.1.1. *Exhaustion of charge carriers*

Exhaustion of charge carriers can be attributed to different phenomena that occur earlier or later according to the section of the wire. By order of appearance:

4.3.1.1.1. Depletion

Depletion is visible very early below 100 nm. It tends to reduce the efficient conduction section (see Figure 4.22) thereby seemingly modifying the conductivity of the wire (its physical section taken as the conduction section):

$$R_{SNW} = \frac{\rho_{eff} L}{s} = \frac{\rho L}{s_{effective}} \qquad [4.21]$$

where ρ, ρ_{eff} are the resistivity of the solid silicon and the resistivity of the nanowire, respectively, s and s_{eff} are the physical section of the wire and the effective section considering the width of depletion, respectively, and L is the length.

For a boron doped nanowire[9], the surface traps generally capture positive charges thereby creating a depletion zone whose width w_d depends on the initial concentration of the boron and the trap surface density. When the depletion width becomes non-negligible compared with the diameter of the wire, its apparent resistivity will increase. This effect is schematized in Figure 4.22. In other words, current is not spread across the whole section of the nanowire, which increases the resistance of the silicon wire, as is shown in Figure 4.22(b). For the nanowire to constitute a NEMS, it must be free to move. After the manufacturing stages that define the 3D geometry, it must be released, i.e. the material located below must be etched (see Figure 4.23(a)). After this stage, it is not unusual to note that the resistance of the nanowire has increased (see Figure 4.23(b)). Measurements have shown that this increase is due to the degradation of the nanowire's surface state. The concentration of impurities capable of trapping charges on the surface increases. The width of the depletion zone becomes increasingly large, reducing in particular the effective conduction section as is shown in Figure 4.23(a). Other effects that can also contribute toward increasing the resistance of the nanowire are the result of the manufacturing process:

9 The majority of carriers are holes.

(1) after release the physical section can be reduced and (2) to a lesser degree the suspended nanowire can present a rougher surface, which reduces the mobility of the charge carriers on the surface. This is a second-order effect as long as conduction is, for the most part, realized in the volume.

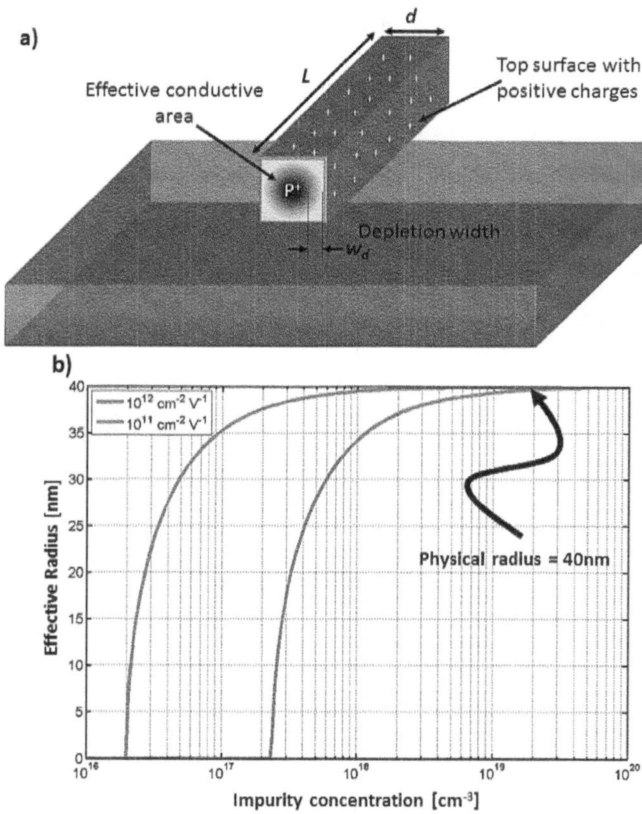

Figure 4.22. *Depletion effect in a square section of a suspended nanowire: a) schematic view of electrical conduction: conduction is confined to the center of the nanowire (dark blue), depleted zone on the periphery (light blue); b) example of calculating conduction radius according to doping for two concentrations of traps on the surface for a 40 nm radius nanowire. For low dopings, it can be seen that the wire becomes insulating. For a color version of the figure, see www.iste.co.uk/duraffourg/nems.zip*

4.3.1.1.2. The deactivation of dopants

This phenomenon tends to reduce the number of free carriers contributing to conduction and constitutes the second observable effect for diameters less than 40 nm. The deactivation mechanism can be explained quite easily by considering that the dopant impurity is a hydrogenoid system [DIA 07] shown in Figure 4.24.

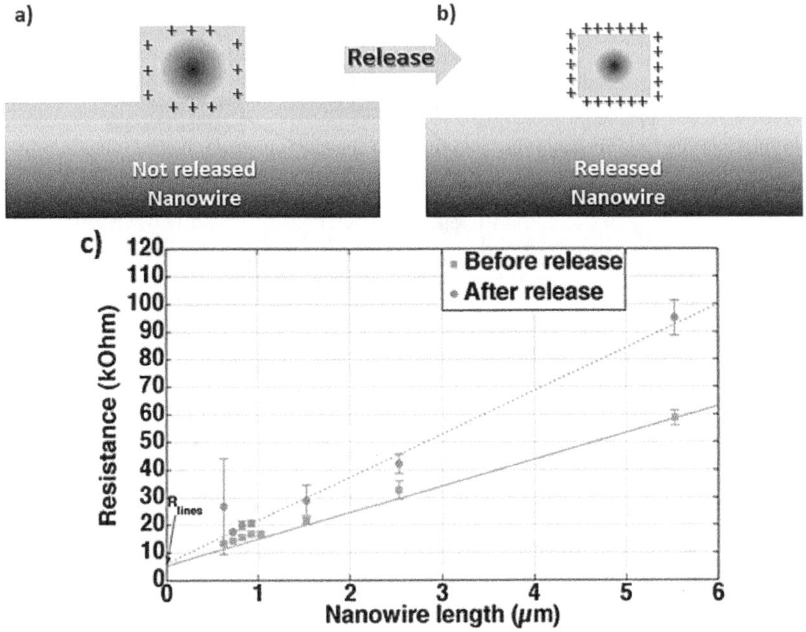

Figure 4.23. *Impact of the release step on depletion: a) conduction section of a nanowire surrounded by silicon dioxide SiO_2 before release; b) conduction section of the same nanowire after release: the concentration of charges on the surface increases; c) resistance measured on a 80 × 160 nm^2 nanowire (doping $10^{19} cm^{-3}$) according to length before and after release: there has been a clear increase in resistance after release. For a color version of the figure, see www.iste.co.uk/duraffourg/nems.zip*

$$E_I^H = \frac{me^4}{2\hbar^2} = 13.6\, eV$$

$$a_B = \frac{\hbar^2}{me^2} = 0.52\, \mathring{A}$$

Figure 4.24. *Hydrogenoid – a_B is the Bohr radius – E_I^H is the link energy*

In a macroscopic case, when this impurity is introduced in a semiconducting crystal, it creates an electrical field that is immediately attenuated by a screening effect due to the displacements of the loads of the crystal around this impurity. The bonding energy (ionization) E_I^0 between the charge brought by the impurity and its core is divided by the permittivity of the semiconductor as is shown by the following phenomenological equations:

$$E_I^0 = \frac{E_I^H}{\varepsilon_{Si}^2}\frac{m^*}{m}$$
$$a_B^0 = a_B \varepsilon_{Si} \frac{m}{m^*} \qquad [4.22]$$

where m, m^* are the mass of a free electron and the effective mass of the electron in the crystal, respectively, and ε_{Si} is the relative permittivity of the silicon ($\varepsilon_{Si} = 11.7$).

As such, a dopant impurity is easily ionized at room temperature. In the case of an environment with very low volume, typically for nanowires whose diameter is approaching 10 times that of the Bohr radius of the boron given in equation [4.22][10], electrostatic screening is reduced. Bonding energy can become much higher than thermal energy k_bT by one order of magnitude. The number of thermally ionized impurities is finally lower, which reduces the number of available charge carriers (holes or electrons). This ionization energy E_I has been calculated in the case of nanowires by ab-initio methods and then approximated by an analytical formula [DIA 07, BJÖ 09]:

$$E_I - E_I^0 \approx \frac{2e^2}{\varepsilon_{Si} d_{eff}} \frac{\varepsilon_{Si} - \varepsilon_{out}}{\varepsilon_{Si} + \varepsilon_{out}} F\left(\frac{\varepsilon_{Si}}{\varepsilon_{out}}\right) \propto \frac{1}{d_{eff}} \qquad [4.23]$$

where F is a function given in [DIA 07]. ε_{out} is the relative permittivity of the environment in which the nanowire is immersed. d_{eff} is the effective diameter calculated by taking depletion effect into account.

10 To be as rigorous as possible, it is necessary to compare to the Debye wavelength which depends on the doping considered (for example, λD~130 nm for 10^{15} cm^{-3} and λD~1.30 nm for 10^{19} cm^{-3}).

Finally, the number of charge carriers is numerically computed by resolving the following equations in turn (iterative algorithm):

$$\begin{aligned} N_d^+ &= N_d \frac{1}{1+2e^{(E_F - E_d)/k_B T}} \\ N_a^- &= N_a \frac{1}{1+1/4e^{(E_a - E_F)/k_B T}} \\ E_{F,a,d} &= f(E_{a,d}, E_{v,c}, T) \end{aligned} \qquad [4.24]$$

where N_d^+ and N_a^- are the concentrations of ionized dopants. E_a, E_d and E_F are, respectively, the ionization energies of acceptor dopants, donor dopants and Fermi energies $E_{F,a,d}$. In the general case, the latter are calculated by the Fermi integral, which is beneficial for approximations as it cuts down on calculation times [AYM 83]. Finally, the concentrations of free carriers are defined by the charge conservation law:

$$\begin{aligned} n &= N_d^+ + p_0 \\ p &= N_a^- + n_0 \end{aligned} \qquad [4.25]$$

where n is the concentration of electrons and p is the concentration in holes (p_0, n_0 are the intrinsic concentrations).

The system of equations [4.22]–[4.25] is resolved by a nonlinear calculation algorithm based on Newton's method. This algorithm integrates both the phenomena of depletion and deactivation. Mobility and resistivity are then calculated based on classical models developed for advanced metal oxide semiconductor field effect transistor (MOSFET) transistors that take into account the main sources of scattering of charge carriers[11]. The whole calculation mode is described in [BJÖ 09].

Numerically resolving equations [4.21]–[4.25] enables the effective conduction radius (conduction in volume) to be calculated according to different physical and morphological parameters (temperature, geometry, doping and surface states). For example, Figure 4.22(b) represents the effective conduction radius according to doping for a nanowire with a 40 nm

11 Interactions: electron/phonon, electron/impurities (Coulomb), electron/surface and electron/impurity on the surface.

diameter, a trap surface density of 10^{11} $cm^{-2}.V^{-1}$ at room temperature. Figure 4.25 represents this effective conduction radius according to the physical radius for three boron doped nanowires at 10^{16} cm^{-3}, 10^{17} cm^{-3} and 10^{18} cm^{-3}, respectively, for a trap surface density of 10^{11} $cm^{-2}.V^{-1}$ and at room temperature. In these two examples, it has been noted that the depletion effect is much more noticeable when doping is low (indeed, the wire doped at 10^{16} cm^{-3} is totally depleted for a radius of 75 nm). Moreover, depletion effect is much stronger when the density of the traps on the surface is high.

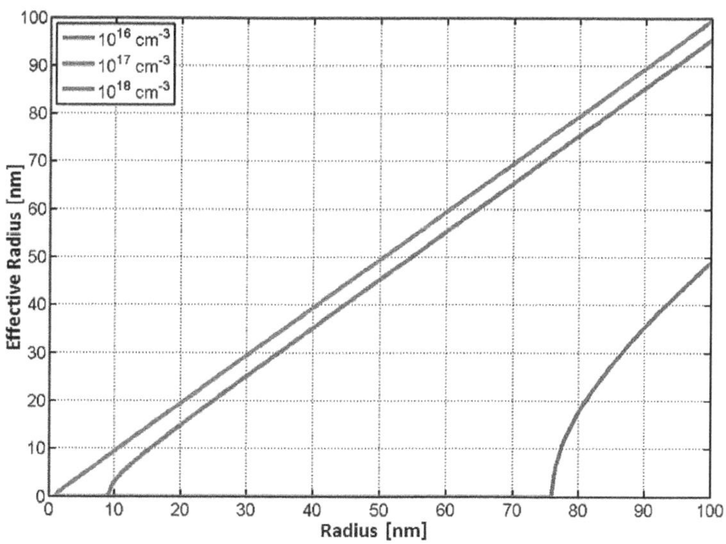

Figure 4.25. *Effective radius according to the physical radius for three boron dopant concentrations – density of the traps on the surface $D_{it}=10^{11}$ $cm^{-2}V^{-1}$ – temperature 300 K. For a color version of the figure, see www.iste.co.uk/duraffourg/nems.zip*

More specifically, Figure 4.22(b) illustrates the fact that the surface state on a 40 nm diameter wire plays an important role in electrical conduction for two extreme cases. The first has a number of traps classically reached by complementary metal oxide semiconductor (CMOS) technology with an appropriate surface treatment (blue curve). The second has a significant number of observable traps and the surface does not undergo any particular treatment, for example, after the release of the silicon wire (purple curve). It has been noted that the wire becomes insulating for much larger radiuses when trap density is high. Surface state plays a crucial role in conduction; it needs to be controlled by treatments that are yet to be defined, but that will probably be inspired by developments in CMOS technology.

Figure 4.26(a) represents the quantity of boron atoms actually ionized according to the physical radius of the nanowire for an initial dopant concentration of 10^{18} cm^{-3}. The dark blue curve represents a wire surrounded by air, whereas the light blue curve represents a wire coated with oxide. In both cases, the quantity of ionized dopants falls with the radius of the wire. This effect does not take into account the "activation", which consists of performing an annealing so that the impurities replace a silicon atom. This process does not have a yield of 100%. It can be stated that coating the wire with a dielectric with strong permittivity reduces the effect of size and constitutes an example of surface treatment. Logically, it can be shown that low room temperatures (for example, 100 K in Figure 4.26(b)) drastically reduce the conduction radius. Indeed, the quantity of ionized dopants bringing free charge carriers is reduced with temperature according to the laws shown in equation [4.24].

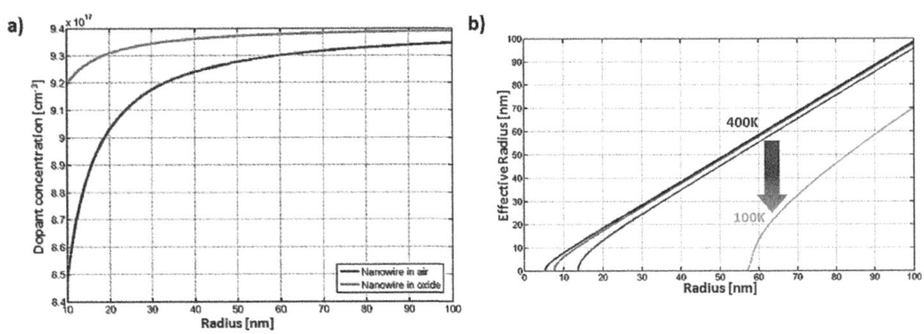

Figure 4.26. *a) Concentration in ionized dopant atoms (boron) according to the physical radius at a room temperature of 300 K (initial dopant concentration of 10^{18} cm^{-3}) for two trap surface concentrations – b) effective conduction radius according to the physical radius for different temperatures (physical radius of 40 nm, initial doping of 10^{18} cm^{-3}). For a color version of the figure, see www.iste.co.uk/duraffourg/nems.zip*

In conclusion, the resistance of the nanowire tends to increase due to surface effects that become more significant. This property can be used to constitute a transistor without junction by varying the surface potential and thereby controlling the width of depletion. The experimental research on this topic cited in Chapter 3 [BAR 14] will be described in detail and will be compared to piezoresistive transduction in the next section.

4.3.1.1.3. Quantum confinement

Quantum confinement is only visible for diameters lower than 5 nm (of the same order as the Bohr radius of the boron). In this case, the structures of the conduction bands are modified as shown in Figure 4.27. For nanowires oriented according to the axes <100> or <110>, optical transition becomes a direct gap transition for diameters lower than 3 nm [HON 08]. The effect of quantum confinement on the exhaustion of charge carriers remains, however, of second order compared to the two other sources, particularly for the diameters considered in our research (> 10 nm), and was not considered in the models developed for the study introduced here.

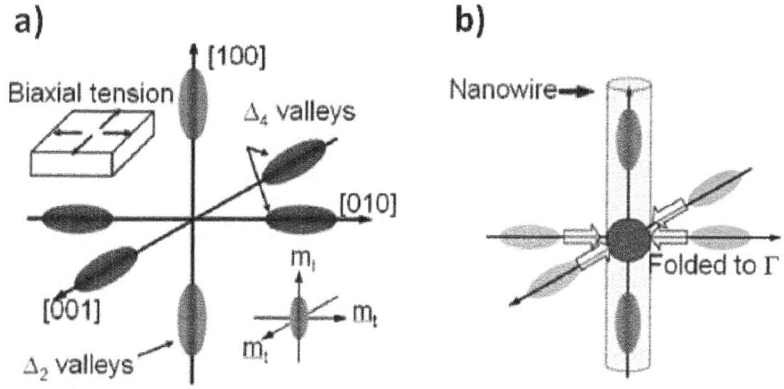

Figure 4.27. *Effect of quantum confinement on conduction bands according to reference [HON 08]: a) iso-energy ellipsoids of silicon; b) effect of confinement of the four ellipsoids according to axes [001] and [010]. For a color version of the figure, see www.iste.co.uk/duraffourg/nems.zip*

4.3.2. *Piezoresistive effect*

4.3.2.1. *Reminder of the theory*

This effect was described in section 2.2.2. To jog your memory, the electrical resistivity of a piezoresistive material varies according to an applied mechanical constraint[12]. The relative resistance variation $\Delta R/R$ is expressed as follows:

$$\frac{\Delta R}{R} = \pi_L \sigma_L = (E\pi_L)\frac{\Delta L}{L} = G\varepsilon_L \quad [4.26]$$

where σ_L is the applied axial constraint. π_L is the axial piezoresistive coefficient linking the resistance variation to the constraint (Pa^{-1}). G is the gauge factor linking the relative resistance variation to the strain (without unit).

In a macroscopic silicon crystal for <100> crystalline orientations, the coefficient G is 100. In the case of P-doped nanowires built using a bottom-up technique, a giant effect has been observed and G can reach 5,000. This scaling effect was not observed in the case of N-doped nanowires. The modification of the electrical properties is particularly due to depletion, described as an electrostatic effect (see section 4.3.1). From the perspective of energy, depletion is induced by the bending of the energy bands at the silicon/air interface [BJÖ 09, SCH 09b]. The constraints applied to the nanowire modify the interface states and indirectly the bending of the bands. As the holes are more affected by this bending change (see section 2.2.2), giant piezoresistivty is exclusively observed in the case of doping P.

To quantify piezoresistivity in nanowires, the previous calculation for their resistivity can be extended to the calculation for the gauge factor G using the model proposed by Rowe [ROW 08] who introduced a modulation parameter of the trap surface density according to the constraint. By then calculating the resistance variation of the wire according to this constraint, the gauge coefficient could be theoretically evaluated for a nanowire placed in the air for different values of trap surface density. The gauge factor for a silicon nanowire with a square section of 40 nm^2 was calculated using this method for two trap surface densities. The results according to doping are presented in Figure 4.28. It can be clearly seen that this giant effect only occurs for very low dopant concentrations. Incidentally, it can be stated that the effect tends toward infinity indicating that the wire becomes insulating (total depletion). At high doping, classic values are expected. Nevertheless,

12 In the most favorable case, this constraint is axial.

if we try to use the nanowire as a sensor, it is clear that a compromise must be found between a high resistivity for a high coefficient and a low resistivity for a reduced coefficient. In other words, this amounts to calculating signal-to-noise ratio by only considering Johnson noise.

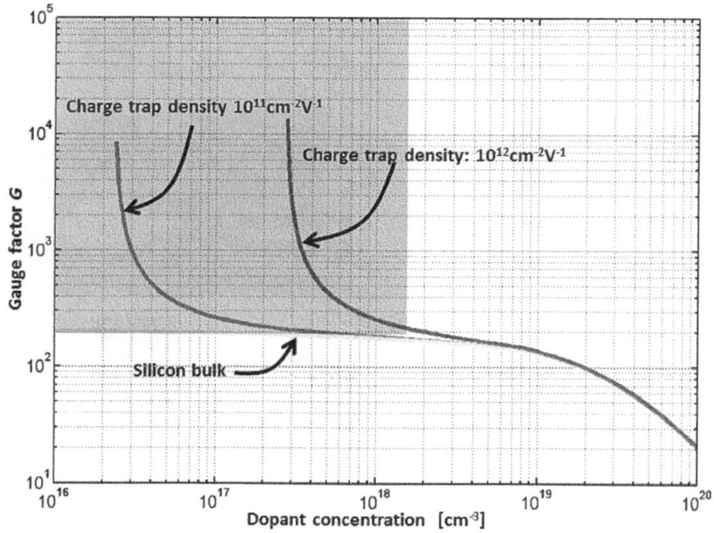

Figure 4.28. *Piezoresistive coefficient G according to the concentration for a doped boron 40 nm nanowire at room temperature T=300 K for trap densities $D_{it}=10^{11}$ $cm^{-2}V^{-1}$ and $D_{it}=10^{12}$ $cm^{-2}V^{-1}$. For a color version of the figure, see www.iste.co.uk/duraffourg/nems.zip*

4.3.2.2. *Preliminary measurement*

The resistance of a nanowire is generally measured according to an applied longitudinal constraint. A silicon strip (with a surface of several cm^2) on which the nanowires have been structured is placed in a constraint bench comprising four "punctual" supports (see the configuration described in Figure 4.29). The strip is deformed (bent) by moving the two central supports while keeping the other two fixed. A photograph of a typical bench is presented in Figure 4.29. For more information, please consult [LUN 04].

The gauge factor is measured from the relative resistance variation according to the constraint. It is then represented according to relation [4.26]. Static current-voltage characteristics are measured for different states of constraints (or strain), enabling relative resistance variation to be deduced according to strain (see Figures 4.30(a) and (b)).

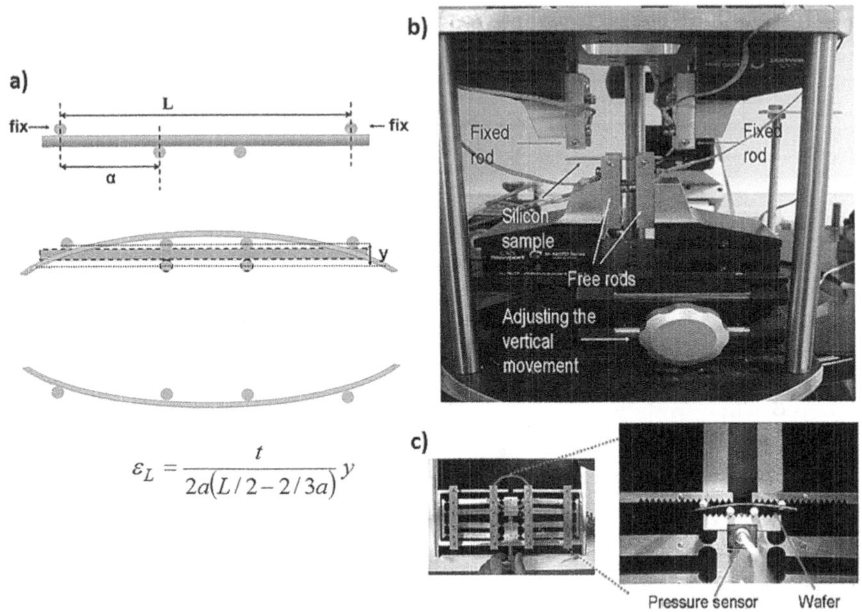

Figure 4.29. *Constraint bench associated with a four point measurement of the resistance – resistance variation is recorded point-by-point according to the elongation ε_L applied: a) ε_L is set by the displacement of the two central mobile supports; b) photograph of an automatic measurement bench that controls the pressure applied and temperature; c) photograph of a manual system*

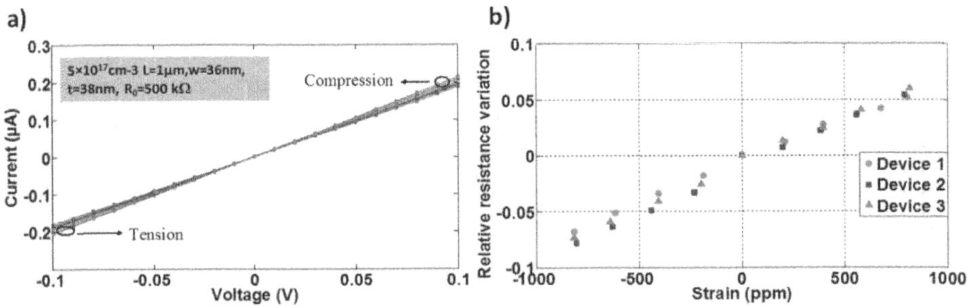

Figure 4.30. *Typical result obtained from low doped wires of low section; a) static characteristics I(V) for different constraints in compression than in voltage; b) relative resistance variation ΔR/R deduced from the curves I(V) according to strain ε_L. For a color version of the figure, see www.iste.co.uk/duraffourg/nems.zip*

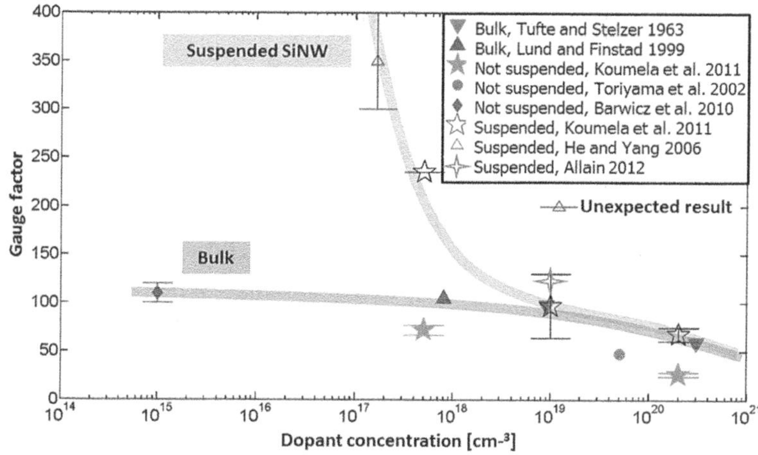

Figure 4.31. *Graph showing the gauge factors measured on solid silicon and on different nanowires (structured with top-down technology via lithography/etching or bottom-up technology by growth); the nanowires have sections varying from 500 to 10^4 nm² and are typically several microns long. Doping is P-type – two clear trends can be identified: the gray curve linking the experimental points obtained for bulk silicon and the green curve linking the experimental points obtained with freely suspended nanowires; the green approximation curve is obtained by calculating the gauge factor on a nanowire of 50 × 50 nm² for a length of 2 μm with a trap surface density of 2×10^{11} cm^{-2}.V^{-1}. It can be seen that the gauge factors measured with fixed nanowires follow a similar trend to those of bulk silicon. For a color version of the figure, see www.iste.co.uk/duraffourg/nems.zip*

Countless measurements of the gauge factor have been made on silicon nanowires, particularly following research conducted by P. Yang's group [HE 06]. Figure 4.31 summarizes the main experiment results obtained over the past 40 years plotting the experimental gauge factor according to dopant concentration. These measurements have been made on freely suspended or fixed (with oxide surrounding the nanowire) nanowires and compared to bulk silicon. Two clear trends emerge: (1) the first trend concerns non-free nanowires. The trend follows a monotonous law that depends on dopant concentration that is similar to the law followed by bulk silicon. The values are a bit lower: this difference can be explained by the transmission conditions of the least efficient constraint in the nanowires while the sample is bent in the measurement bench; (2) and the second trend groups together all the free nanowires. The trend follows a monotonous law with abrupt growth when dopant concentration is reduced. This experimental law is very close to the law we calculated for the theoretical model previously presented. The high concentration experiment point (10^{19} cm^{-3}) remains, however, to be explained. This measurement corresponds to that recorded by P. Yang's group [HE 06] for nanowires manufactured by growth (bottom-up process).

We have seen that the surface state influences depletion width, the resistance of the nanowire and consequently its piezoresistivity. In other words, surface charges cause a difference in potential between the surface and the center of the nanowire, which induces a bending of the energy bands at the silicon/air interface. It is, nevertheless, possible to control this surface potential using an electrode close to the nanowire. When the wire is put in movement, the electrostatic capacitance formed by the latter changes and consequently the surface potential changes. The current passing through the wire will thus be modulated according to the mechanical movement. In so doing, the nanowire is equivalent to a suspended transistor without junction.

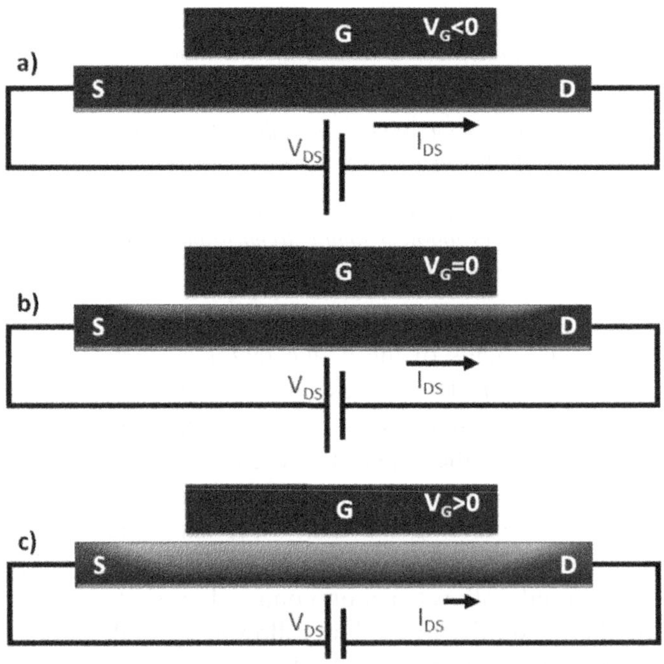

Figure 4.32. *Operating regime of a P-doped (boron) nanowire without junction; a) when the gate voltage applied to the electrode is negative, the nanowire is on (ON-state). In addition to an applied voltage compensating the exact potential induced by surface charges, the wire is switched on along its entire physical width; b) for null voltage, the nanowire is on with the exception of the residual depletion width induced by surface charges; c) for positive voltage, the wire is depleted until what is known as pinch-off voltage, when it switches off (OFF-state). For a color version of the figure, see www.iste.co.uk/duraffourg/nems.zip*

To conclude the section, this detection principle, inherent to the scaling effect, will be compared with piezoresistive detection.

To this end, we will take another look at the P-doped nanowire described in Figure 2.22. The different operating regimes are described in Figure 4.32. Nanowires configured as transistors without junction can control resistance using a simple gate electrode (or two electrodes placed on either side of the wire). To obtain this effect, electrostatic gaps must be very narrow, typically lower than 100 nm. In this case, the nanowire is on when gate voltage is null (see Figure 4.32(b)). Reducing voltage to negative values, reduces depletion width until the wire switches on completely (see Figure 4.32(a)). At this value, known as flat band voltage V_{FB}, the wire conducts over its entire physical section. When $V_G<V_{FB}$, an accumulation current superimposes itself on the volume current. By increasing voltage $V_G>0$ depletion width increases, reducing the drain current until the conduction channel pinches-off (see Figure 4.32(c)), at this stage, $V_G=V_p$, pinch-off voltage. When $V_G>V_p$, the nanowire is turned off. The current passing through the nanowire can be modeled according to V_G, V_{DS} as well as V_{FB} depending on the surface charge density. Supposing abrupt depletion [COL 90], the current circulating in the volume of the nanowire (without considering an accumulation at high voltage) can be expressed as follows [COL 10, KOU 13]:

$$I_{DS}(V_G,V_{DS})=\frac{2t}{\rho L}\left[\left(\frac{w}{2}-\frac{C_i}{eN_a}(V_G-V_{FB})\right)V_{DS}+\frac{C_i}{eN_a}\frac{V_{DS}^2}{2}\right]$$
$$g_m=\frac{\partial I_{DS}}{\partial V_G}=\frac{-2t}{\rho L}\frac{C_i}{eN_a}V_{DS} \qquad [4.27]$$

where t, w, L are the thickness, the width and the length of the nanowire, respectively. ρ is the effective resistivity of the nanowire (see equation [4.21]). C_i is the capacitance between the nanowire and the gate electrode including a potential oxide layer. N_a is the concentration of dopants. g_m is the transconductance of the wire.

Figure 4.33 shows the static characteristics. Specifically, Figure 4.33(b) shows the drain current according to gate voltage. For low drain voltages (here V_{DS} = –1.5 V), the model shows a favorable trend and enables the current to be evaluated with a good level of precision up until pinch-off voltage, which can be estimated at the current bend point, i.e. V_p~20 V. By increasing V_{DS} (in the example $V_{DS} = -4.5$ V), theory (equation [4.27]) is no longer satisfactory as it supposes that the mobility of the charge carriers (holes) is constant, which is not the case at high transverse or longitudinal fields. Moreover, the actual length of the channel formed by the nanowire and the series resistance depend on V_{DS}. In simple terms, the drain and source zones which are supposed to be inexhaustible charge reservoirs see their dimensions vary with V_{DS}. This effect is also visible on the transconductance g_m presented in Figure 4.33(c).

We will now compare the piezoresistive transduction presented in section 2.2.2 (see Figure 2.24) and this new detection based on the junctionless-transistor. We will first look to define the expression of the motional current (induced by movement) according to vibration amplitude using an approach similar to that presented to define this current for piezoresistive detection (see equations [2.27] and [2.28] in section 2.2.2). To do so, let us start from the expression of transconductance given in equation [4.27] which lets us define drain current according to gate voltage. At small signal, the latter varies around a constant value V_G with an amplitude δV_G. In the case studied, δV_G depends on the capacitance bridge formed by the air capacitance that varies with the vibration and the oxide capacitance surrounding the nanowire:

$$I_{ds}(\omega) = g_m \delta V_G(\omega) \approx g_m V_G \frac{\delta C(\omega)}{C_{gap}} = g_m V_G \frac{y(L/2,\omega)}{g} \qquad [4.28]$$

where $y(L/2, \omega)$ is the vibration amplitude at the center of the nanowire, defined in section 2.1 (see equation [2.2]). The oxide capacitance is negligible in relation to the capacitance formed by the air gap in series. The capacitance C_i is, therefore, reduced to C_{gap}. Equation [4.28] shows that signal is directly proportional to the vibration amplitude. In the case of piezoresistive detection, current follows quadratic law with displacement. Like for our study on piezoresistive detection, we will consider the nanowire in Figure 2.22 actuated by the same electrostatic force. As the current is proportional to displacement, the frequency of the drain current from detection by junctionless-transistor is identical to actuation frequency. On the other hand, due to quadratic law, the frequency of the so-called "piezoresistive" current is double that of actuation frequency. On this same wire, the difference between piezoresistive current and current from junctionless-transistor can be identified.

Let us return to the "down-mixing" scheme in which the nanowire is polarized by a voltage V_{DS} and actuated by a voltage V_G, which is the sum of a continuous component V_{DC} and a sinusoidal component $V_{AC}(\omega)$. We are, therefore, looking to measure the electromechanical drain current I_{DS} by modulating the readout signal V_{DS} to a frequency equal to $\omega \pm \Delta\omega$ and the excitation signal at frequency ω centered on ω_0 (see Figure 4.34 and Figure 2.23). By picking up the expression $y(L/2,\omega) = y_1(\omega)\phi(L/2)$ at resonance ω_0 knowing that $V_G=V_{DC}+V_{AC}(\omega)$, the electromechanical (or motional) drain current (peak) is:

$$I_{DS}(\omega) = g_m V_{DC}\phi\left(\frac{L}{2}\right)\frac{(C_n\varepsilon_0 Lt)V_{DC}V_{AC}(\omega)}{g^2}\frac{1}{m_{eff}\left(\omega_0^2-\omega^2+j\omega\omega_0/Q\right)} \qquad [4.29]$$

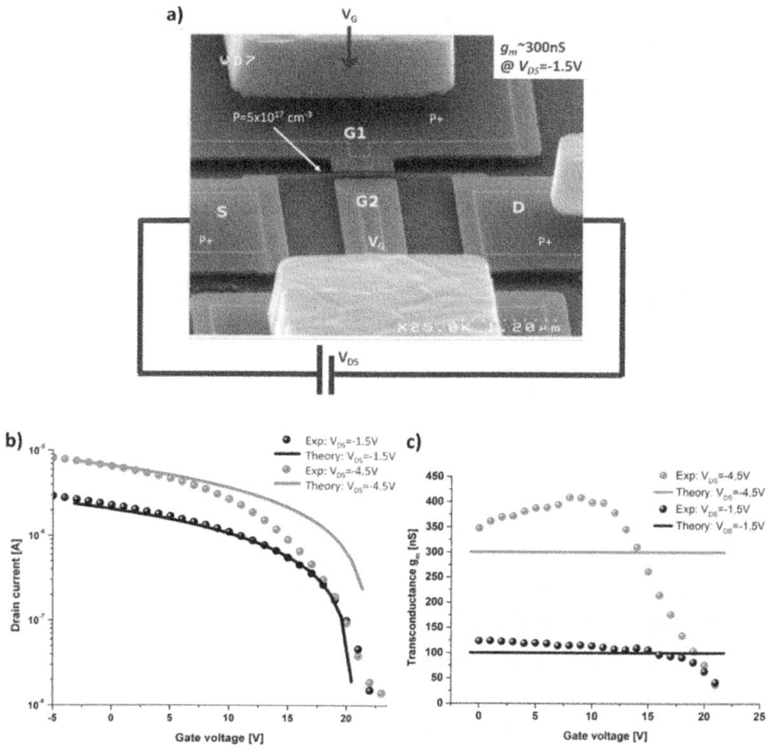

Figure 4.33. *Static characteristics: symbols, measurements and traits [4.27]. a) Polarization of a nanowire; b) drain current according to gate voltage for two drain voltages: pinch-off voltage is about 20 V and flat band voltage (not visible in the measurement) is estimated at 3 V; c) transconductance g_m according to gate voltage (adapted from [KOU 13])*

At resonance and by expressing the effective mass and transconductance, we get an expression of electromechanical current according to the applied voltages and the morphological parameters of the nanowire:

$$I_{DS}(\omega_0) = g_m V_{DC} \frac{4}{3}\frac{(C_n\varepsilon_0 Lt)V_{DC}V_{AC}(\omega_0)}{g^2}\frac{Q}{m\omega_0^2}$$
$$I_{DS}(\omega_0) = \frac{-2t}{\rho L}\frac{C_i}{eN_a}V_{DS}(\omega_0 - \Delta\omega)V_{DC}\frac{4}{3}\frac{(C_n\varepsilon_0 Lt)V_{DC}V_{AC}(\omega_0)}{g^2}\frac{Q}{m\omega_0^2} \quad [4.30]$$
$$I_{DS}(\omega) \propto V_{DS}(\omega_0 - \Delta\omega)\times V_{AC}(\omega_0)$$

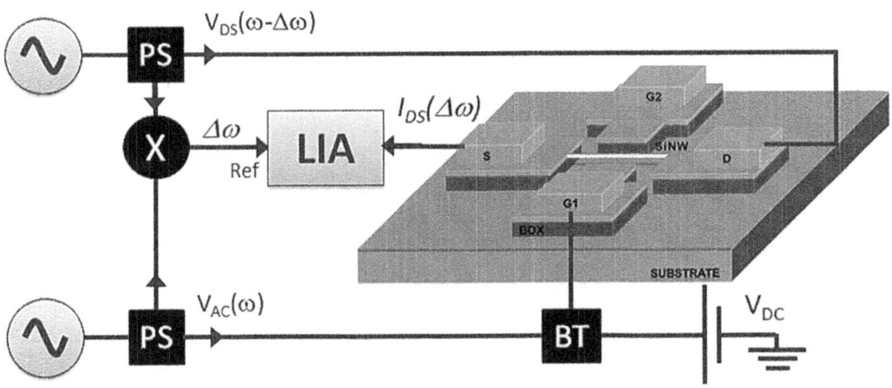

Figure 4.34. *Readout scheme by "down-mixing", the nanowire vibrating at high frequency: V_{DS} is the readout voltage whose frequency varies at $\omega \pm \Delta\omega$ in an interval centered on ω_0; V_{AC} is the actuation signal whose frequency varies at ω. As the nanowire behaves like an Rf mixer [4.30], the output current I_{DS} varies at frequency $\Delta\omega$ and will be detected by a lock-in amplifier (LIA)*

All that remains to be done is to compare the electromechanical drain current obtained by piezoresistive detection (see section 2.2.2 and Figure 2.24) with the current from detection by junctionless-transistor. The experimental curves are presented in Figure 4.35. For clarity, the results obtained by piezoresistive detection are shown on the side. The linear laws for detection by junctionless-transistor and the quadratic laws for piezoresistive detection were experimentally verified. The theoretical curves correspond with the measurements. Piezoresistive transduction in measurement conditions remains the most efficient. To find out more about this comparison, please read article [KOU 13]. Having said that, by reducing the distance between the wire and the gate, junctionless-transistor detection could be just as efficient. One of its advantages is that it is, in essence, linear. Other research has, in fact, proven this detection to be very promising as it has different profiles of doping and electrostatic gaps [BAR 14].

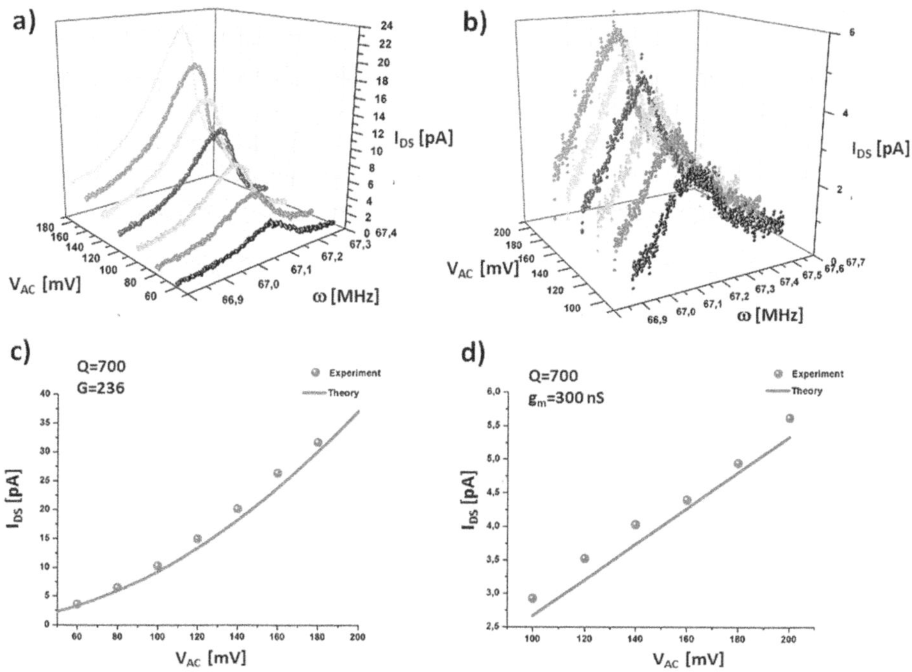

Figure 4.35. *Comparison between piezoresistive detection and detection by junctionless-transistor with V_{DC}=300 mV and V_{DS}=70 mV: a) electromechanical current I_{DS} according to frequency and actuation voltage for piezoresistive detection; b) electromechanical current I_{DS} according to frequency and actuation voltage for detection by junctionless-transistor; c) piezoresistive detection: current at resonance according to actuation voltage and theoretical approximation from equation [2.38] considering a gauge factor of 236 (from the model developed in section 2.2.2) and a quality factor Q=700; d) detection by junctionless-transistor: current at resonance according to actuation voltage and theoretical approximation from equation [4.30] considering a transconductance g_m of 300 nS (static measurement) and a quality factor Q=700. For a color version of the figure, see www.iste.co.uk/duraffourg/nems.zip*

4.4. Optomechanical nano-oscillators and quantum optomechanics

We have described several examples of scaling effects and presented NEMS-specific transductions. Current research is increasingly focusing on applications. Conclusions drawn by these studies predict that Casimir energy will be used to actuate micro- or nanosystems. If the charge density at the surface of interfacing mechanical elements is sufficiently low, or if it can be offset, it has been shown that the Casimir force can actuate torsion micromirrors [CAP 01]. In essence, the Casimir force is attractive when two interfacing planes are flat and not very rough. It is nevertheless possible to make this force repulsive by nanostructuring the surfaces or by using metamaterials. The sign of the force, therefore, depends on the distance separating the planes. Let us imagine that the force periodically inverses according to the variation of the gap between the suspended element and a fixed plane [ROD 11]. In these conditions, the energy from the quantum fluctuations of the vacuum can be used to ensure self-oscillation.

Naturally, we can try to generalize this approach by using the energy of the photons stored in an optical cavity like a Fabry–Pérot system including a mobile mirror (see Figure 4.36(a)). The mirror is put in movement by transferring the momentum of the photons when they are reflected on its surface (radiation pressure). In response, the movement of the mirror modifies the optical properties of the cavity, thus enabling displacement to be measured. This optomechanical transduction enables extremely fine resolutions, ranging from 10^{-20}m/√Hz to 10^{-16}m/√Hz[44], to be reached. To understand the mechanisms put in play, we will describe the properties of the Fabry–Pérot cavity constituting an easy-to-understand optical resonator. Its spectral response in transmission *FP(ν)* is an Airy function comb[13]. The separation between two consecutive peaks is called the free spectral range (FSR). The Airy function is written as follows:

$$FP(\nu) = \frac{T_{max}}{1+M\sin^2(\varphi/2)} \qquad [4.31]$$

With: $T_{max} = \frac{T^2}{(1-R)^2} \sim T^2$ for T<<1, $M = \frac{4F^2}{\pi^2}$, $\varphi = \frac{4\pi d}{\lambda} = \frac{4\pi d\nu}{c}$ (F is the finesse of the peaks $F = \frac{ISL}{\delta\upsilon}$).

13 In the case of reflection, the transfer function *R* is such that $T+R=1$.

Transmission is maximal for any cavity frequency ν_c such that:

$$\nu_c = \frac{kc}{2d}; k \in \mathbb{N}$$
$$ISL = \frac{c}{2d} \qquad [4.32]$$

The eigen-frequencies of the cavity ν_c are, therefore, all proportional to $c/2d$ (c is the speed of light in free propagation, d is the separation distance). It is interesting to define the quality factor of the optical cavity $Q_{opt} = \nu_c/\delta\nu$, which is vitally important in optomechanical detection, as will be seen later. This estimates the capacitance of the cavity to store light energy, in other words, it enables us to find out the lifetime of a photon in the cavity $\tau_{phot} = Q/\nu_c$.

When one of the mirrors moves by a quantity δx, the length of the cavity changes, inducing a variation in resonance frequency ν_c (see Figure 4.36). Normally, an optomechanical coupling coefficient g_x is introduced to characterize coupling efficiency:

$$\nu = \nu_c + g_x\delta x \qquad [4.33]$$

where g_x is the derivative of the eigen-frequency in relation to displacement $\frac{\partial \nu}{\partial(\delta x)}$. In the simple case of a Fabry–Pérot cavity, $g_x = -\nu_c/d$. Resonant optical cavities such as disks, rings and photonic crystals behave in a similar way. The general expression of g_x remains true but takes on a more complex form.

The two readout principles are described in Figure 4.37. In the first principle, the frequency of the laser (its wavelength) is detuned compared to the resonance frequency of the cavity (see Figure 4.37(a)). As is shown in Figure 4.36, the transfer function of the cavity varies according to the movement of the mirror, thereby modulating the intensity of the light transmitted and reflected because the operating point is situated in the slope of the transfer function. Input-output phase shift is modulated very little because the operating point is located on a plateau. In the second approach, the frequency of the laser matches the resonance frequency of the cavity. In this case, the modulation of intensity is almost null around the maximum of the transfer function. On the other hand, the operating point is located in the abrupt slope of the phase change. The phase shift of the optical signal is, therefore, modulated by the movement of the mobile mirror.

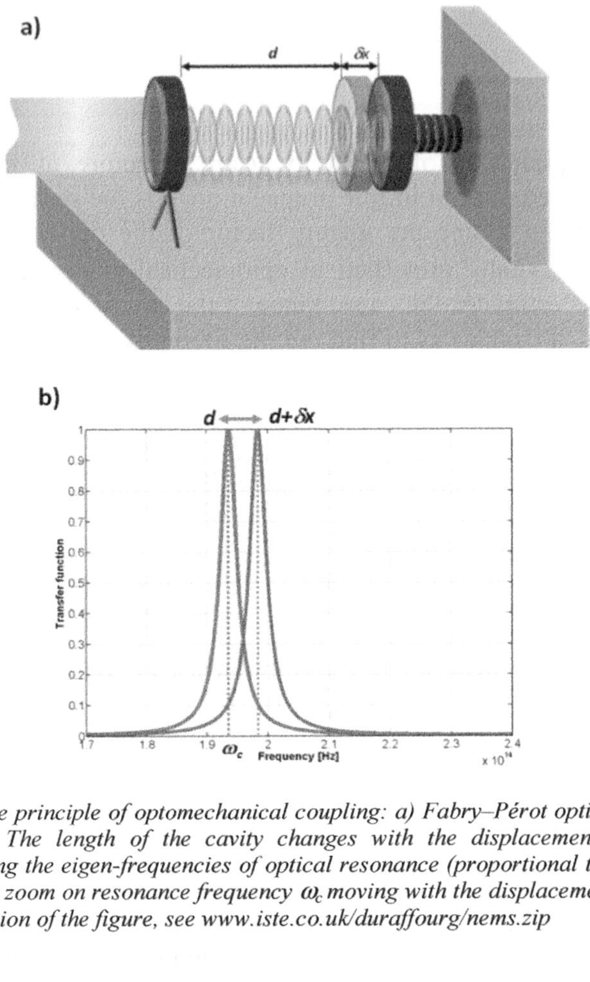

Figure 4.36. *The principle of optomechanical coupling: a) Fabry–Pérot optical cavity with a mobile mirror. The length of the cavity changes with the displacement of the mirror thereby modifying the eigen-frequencies of optical resonance (proportional to c/2(d+δx) (see Figure 4.11); b) zoom on resonance frequency ω_c moving with the displacement of the mirror. For a color version of the figure, see www.iste.co.uk/duraffourg/nems.zip*

It is easy to define the relation between the relative variation in transmitted or reflected optical power $\Delta P/P_0$ and the displacement of the mirror δx using equations [4.31] and [4.33]. This expression can define the important parameters for optimizing resolution limit in displacement:

$$\frac{\delta P}{P_0} = \left[\left(\frac{\partial FP}{\partial \nu}\right)\left(\frac{\partial \nu}{\partial x}\right)\right]\delta x$$
$$\frac{\delta P}{P_0} \propto g_x \cdot \frac{Q_{opt}}{\nu_c} \cdot \delta x \qquad [4.34]$$

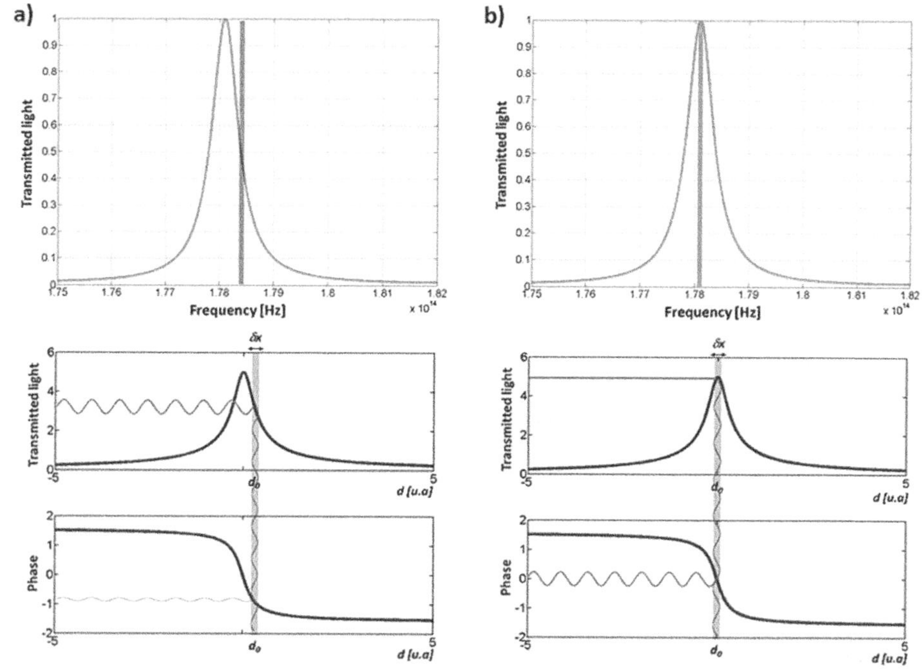

Figure 4.37. *The principle of optomechanical detection: a) probe laser enabling movement to be read is detuned compared to the resonance eigen-frequency of the optical cavity: a variation in the position of the mirror causes the intensity to be modulated and a low variation in the transmitted phase signal (likewise for the reflected signal); b) probe laser enabling movement to be read matches the resonance eigen-frequency of the optical cavity: a variation in the position of the mirror causes considerable phase modulation and very low variation in intensity of the transmitted signal (likewise for the reflected signal). For a color version of the figure, see www.iste.co.uk/duraffourg/nems.zip*

It is therefore necessary to inject lots of light P_0 and have a high quality factor Q_{opt} and strong electromechanical coupling. A great amount of research has been conducted to optimize the quality factor of optical microresonators. The latter can be ring waveguides, microdisk, micro toroidal waveguides in which the light turns or photonic crystals. All these systems present an optical response similar to the transfer function defined for Fabry–Pérot[14] (see equation [4.31]). They are

14 A ring can be considered as a system of *N* mirrors, *N* tending toward infinity.

schematized in Figure 4.38; the typical quality factors and their volumes are provided.

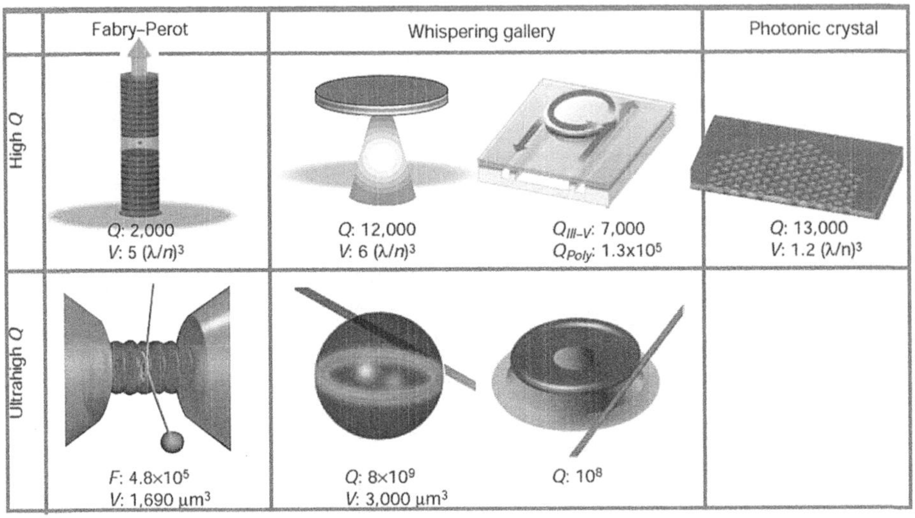

Figure 4.38. *Different types of optical resonators (volumes and quality factor) – schemes from [VAH 03]. For a color version of the figure, see www.iste.co.uk/duraffourg/nems.zip*

Three types of optomechanical systems have been invented using optical microresonators. They are schematized in Figure 4.39. The first approach corresponds to the Fabry–Pérot resonator. The second approach consists of placing a mechanical element, such as a beam or membrane, close to a fixed optical microresonator. The effective refraction index of the waveguide forming the optical microresonator, therefore, changes according to the distance between the mechanical element and the guide (see Figure 4.39(b)). The mechanical element is generally placed several hundred nanometers from the optical resonator. In doing so, the optical resonance frequency changes. In the third approach, the optical resonator is suspended and itself constitutes the mechanical element in vibration (see Figure 4.39(c)). The breathing movement of the optomechanical resonator modulates the optical resonance frequency at the mechanical resonance frequency. This approach is equivalent to Fabry–Pérot's approach with a mobile mirror as it is based on the variation in the length of a photon's mean path. To get high optomechanical couplings, high-frequency volume deformation modes should preferably be used.

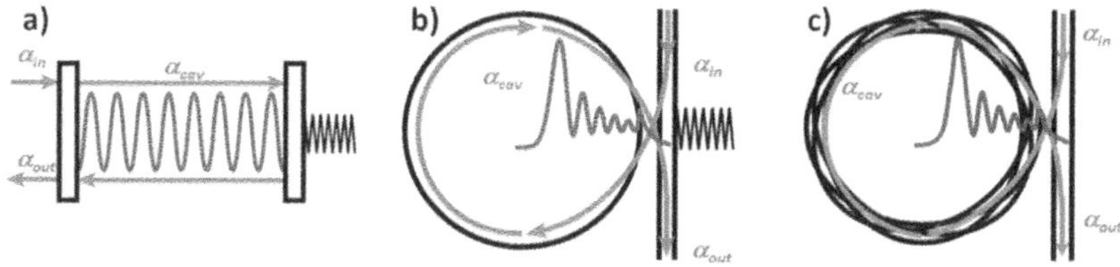

Figure 4.39. *The three types of optomechanical couplings (adapted from [EIC 07]): a) Fabry–Pérot cavity whose length varies with the movement of the mobile mirror; b) microresonator coupled with a suspended element (guide, cantilever, etc.): the optical frequency of the microresonator varies with the movement of the mechanical system or the right guide positioned in near field, by modifying the effective index of the waveguide; c) suspended microresonator itself constituting the mechanical resonator: the length of the optical cavity varies according to the mechanical deformations of the resonator. For a color version of the figure, see www.iste.co.uk/duraffourg/nems.zip*

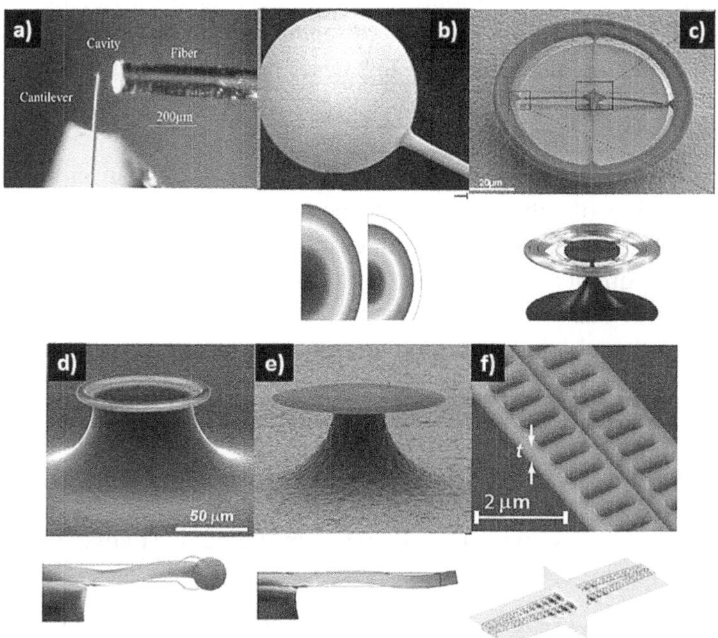

Figure 4.40. *Examples of resonant optical cavities similar to those described in Figure 4.38: a) Fabry–Pérot cavity realized between a cantilever and an optical fiber (image adapted from [MET 08]); b) microsphere (see [CAR 07]); c) optical microtoroidal guide supported by mechanical arms (according to [ANE 08]); d) optical microtoroidal guide vibrating at "Whispering Gallery Mode" (according to [SCH 06]); e) optical microdisk vibrating at "Whispering Gallery Mode" (according to [SRI 11]); f) coupled photonic crystals (according to [EIC 09]). For a color version of the figure, see www.iste.co.uk/duraffourg/nems.zip*

Examples of the realization of these optomechanical systems are shown in Figure 4.40. Figure 4.40(a) is a Fabry–Pérot cavity whose mobile mirror is a cantilever; the fixed mirror is an optical fiber. Figures 4.40(b)–(e) are different suspended optical cavities that are spherical, toric or disk-shaped, respectively. These systems constitute optical resonators operating in "Whispering Gallery Mode" (WGM) [MAT 06, ILC 06] and correspond to the transduction scheme shown in Figure 4.39(c) and Figure 4.41. The frequencies of mechanical deformation are at the order of the GHz. Figure 4.40(f) corresponds to an optical cavity formed by photonic crystals that deform according to the high-frequency volume mode at around 1 GHz. Their relative movement induces a variation in the optical resonance frequency of the set.

For mobile mirrors, it is intuitively understood that the incident photons exchange a certain linear momentum. If there is a sufficient number of protons and if the energy brought by each photon is sufficient, radiating pressure will actuate the mirror. In this case, optical force is written as:

$$F_{rad} = 2\hbar k\bar{I} \quad [4.35]$$

where k is the wave vector of the photons, $2\hbar k$ is the mean momentum. $\bar{I}$ is the mean number of intra-cavity photons per second (mean flow of photons in the cavity) and depends on the finesse or the quality factor of the cavity (the higher the quality factor of the cavity, the longer the lifetime of an intra-cavity photon). The average number of photons present in the cavity per second is just as high. Radiation pressure is proportional to the quality factor.

We have discussed optomechanical detection, however, the systems presented above are actuated by an optical force whose origin differs according to the architectures. Let us consider the case of optical cavities whose length varies with mechanical movement (Fabry–Pérot, WGM systems, see Figure 4.41).

For exercise purposes, it is interesting to define the expression of optical force due to radiation pressure by writing the Hamiltonian of the optomechanical system:

$$\begin{aligned} & \widehat{H}_{opt} + \widehat{H}_m \\ \widehat{H}_{sys} &= \hbar\omega(t)\left(\hat{a}^\dagger\hat{a} + \tfrac{1}{2}\right) + \hbar\omega_m\left(\hat{b}^\dagger\hat{b} + \tfrac{1}{2}\right) \\ & \hbar\omega_c\left(\hat{a}^\dagger\hat{a} + \tfrac{1}{2}\right) + \hbar\omega_m\left(\hat{b}^\dagger\hat{b} + \tfrac{1}{2}\right) + \hbar g_x\hat{x}\hat{a}^\dagger\hat{a} \end{aligned} \quad [4.36]$$

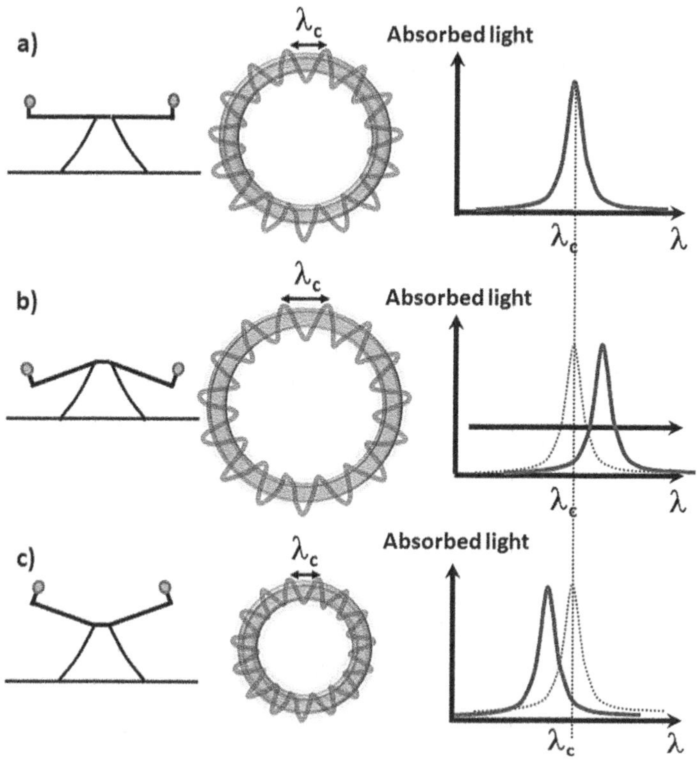

Figure 4.41. *Operating mode of a WGM toroidal resonator – the blue line represents the guided optical field (scheme inspired by T. Carmon's workshop FRISNO-8, see [CAR 07]): a) resonator at rest without actuation with an optical resonance at wavelength λ_c; b) the resonator is actuated at its mechanical resonance frequency by the pressure of the photons circulating in the waveguide, causing an extensional deformation: when the optical cavity extends, the resonance wavelength increases; c) when the optical cavity retracts, the resonance wavelength decreases. For a color version of the figure, see www.iste.co.uk/duraffourg/nems.zip*

In this representation, electromagnetic energy is quantified and appears in the form of an operator $\widehat{H}_{opt}$. $\hat{a}^{\dagger}$ is the photon creation operator and $\hat{a}$ is the photon annihilation operator. The product $\hat{a}^{\dagger}\hat{a} = \hat{n}$ corresponds to the intra-cavity photon number operator carrying quantum energy $\hbar\omega(t)$. Likewise, the vibrations of the mirror can be quantified in the form of quanta energy known as phonon. $\hat{b}^{\dagger}$ and $\hat{b}$ are the phonon creation and annihilation operators, respectively. If you would like to find out more about this quantum approach, there are countless works to read, such as the book written by Braginsky [BRA 85]. Using equation [4.31], it is easy to deduce the optomechanical interaction Hamiltonian $\widehat{H}_{int}$ and the resulting radiation pressure force:

$$\widehat{H}_{int} = \hbar g_x \hat{x} \hat{a}^{\dagger} \hat{a} \qquad [4.37]$$

$$\widehat{F}_{rad} = -\frac{\partial \widehat{H}_{int}}{\partial \hat{x}} = -\hbar g_x \left(\hat{n} + \frac{1}{2}\right) = \frac{\hbar \omega_c}{d}\left(\hat{n} + \frac{1}{2}\right) \qquad [4.38]$$

It should be noted that the radiation force written in this form depends not only on the number of intra-cavity photons, but also on the energy of the vacuum fluctuations in the cavity through the quantity $\hbar\omega_c/2$. This term is, in fact, the intra-cavity contribution to the Casimir force (discussed previously). By rewriting radiation force according to mean light intensity $\hat{I} = \hat{n}/\tau_{phot}$, we get the same expression expected Q that was not included in the previous expression:

$$\widehat{F}_{rad} = \frac{2\hbar\omega_c}{c} Q\hat{I} = 2\hbar k Q\hat{I} \qquad [4.39]$$

Radiation pressure is, therefore, proportional to the mean photon flow $\hat{I}$. The presence of the quality factor Q demonstrates the fact that the photons make a large number of round trips in the cavity. As a guide, without optical resonance, a 1 W laser beam corresponds to a mean photon flow of 10^{18} photons/s, i.e. a mean force of the order of just a few nano-newtons. This rough estimation shows that radiation force remains quite modest. On the contrary, it becomes significant when the cavity has a high quality factor. In the case of a WGM acoustic-optical resonator with a quality factor ranging from $Q_{opt}{\sim}10^5$ to 10^6, the photons make countless circuits of the circular guide and exchange a certain linear momentum with their optical edges, causing the structure to deform slightly.

Another optical force of dipolar origin [VAN 10] is also put in play when two guides, or a guide and a mechanical element, are coupled (separated by a distance of 100 nm). This force derives from the electromagnetic energy present between the two systems (associated with the evanescent field leaking outside the guide) as is shown in Figure 4.42. This optical force was calculated by Povinelli *et al.* [POV 05, MA 12]. As the variation in energy is very abrupt, the resulting dipolar force can be large enough to vibrate a mechanical structure with an amplitude of several nanometers.

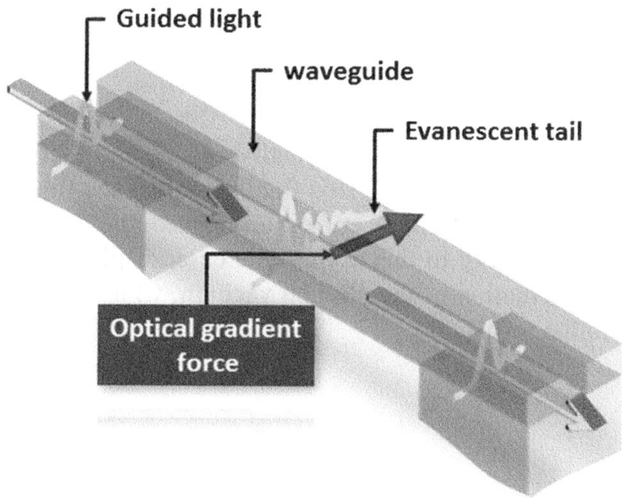

Figure 4.42. *Optical dipolar force: the evanescent field originating from the guide, in which light has been injected, is coupled in the second guide. This coupling is at the origin of optical dipolar force*

Let us imagine that the inverse of the resonance frequency of the mechanical resonator is of the same order of magnitude as the mean lifetime of the photons in the cavity. In the case in point, the photons and phonons can interact. In other words, optical force can be used as a damping force for mechanical resonance or inversely as an amplifying force for movement. This interaction is commonly called "back-action" [KIP 08, SCH 09]: the mechanics have an effect on the optics and in return the optics have an effect on the mechanics (see Figure 4.43). More specifically, in a strong coupling regime, when the laser does not correspond with the resonance frequency of the optical cavity of a quantity exactly equal to the mechanical resonance frequency (toward high frequencies), the incident photons give up energy to the mechanical resonator by a Stokes process[15]. Inversely, if the frequency of the laser shifts toward low frequencies, the photons absorb the energy of the mechanical resonator by an anti-Stokes process. The latter is, therefore, cooled by optical interaction. At the same time, the gradient of the optical force also acts as an optical spring, causing a slight shift in mechanical

15 A Stokes process is an inelastic collision inducing an exchange of a linear momentum.

resonance frequency toward low frequencies, like for the softening effect induced by electrostatic actuation force discussed in this book [BAN 00] (see Chapter 2, section 2.2.1).

The cooling effect is used to drastically reduce the thermal Brownian noise of the mechanical resonator. This finally enables the fundamental state of the resonator to be reached, corresponding to the zero-phonon state (T = 0) [ASP 13, CLE 09]. In this asymptotic case, displacement noise is set by the quantum fluctuation of the phonon vacuum, which is the fundamental limit intrinsic to the resonator.

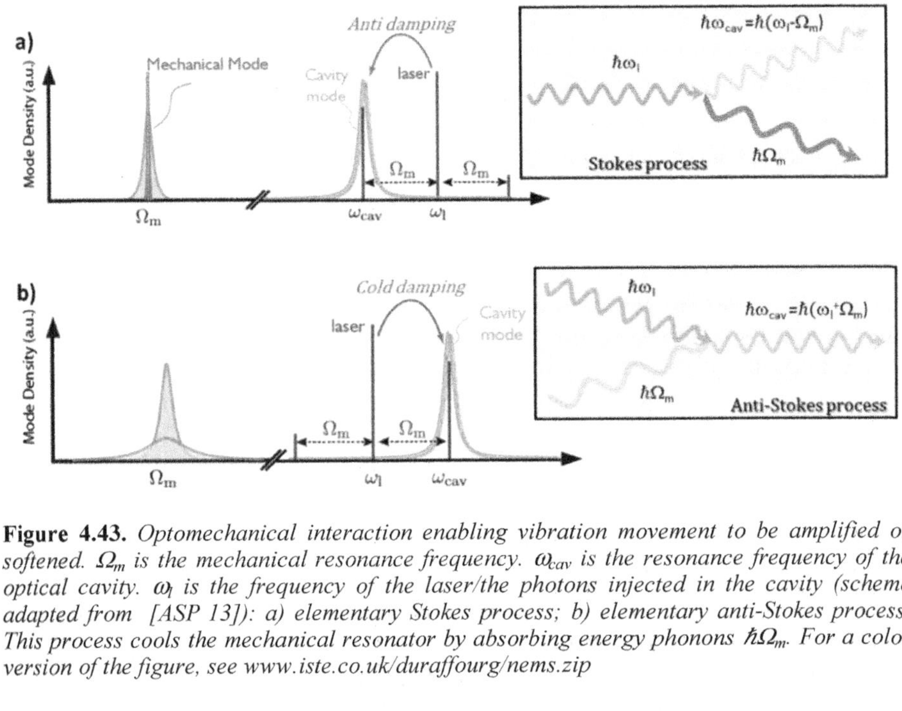

Figure 4.43. *Optomechanical interaction enabling vibration movement to be amplified or softened. Ω_m is the mechanical resonance frequency. ω_{cav} is the resonance frequency of the optical cavity. ω_l is the frequency of the laser/the photons injected in the cavity (scheme adapted from [ASP 13]): a) elementary Stokes process; b) elementary anti-Stokes process. This process cools the mechanical resonator by absorbing energy phonons $\hbar\Omega_m$. For a color version of the figure, see www.iste.co.uk/duraffourg/nems.zip*

On a more practical level and to conclude this fundamental research, it is possible to imagine realizing an entirely optical self-oscillating system that uses the principle scheme described in Chapter 2 (section 2.3, Figure 2.29) with an optomechanical resonator using one of the optical forces and optical detection [HOS 10].

4.5. Conclusion

This chapter gives an overview of scaling effects in NEMS and makes no claim of providing an exhaustive study of the physical phenomena originating from the reduction of size. The thermoelectrical properties of silicon nanostructures or nanostructures made from other materials, for instance, are very promising and deserve to be discussed in their own right. To find out more about thermal and thermoelectrical properties in structures with small dimensions, please refer to [DRE 07] and [CAH 03]. Similarly, "cool" NEMS, known as quantum electromechanical systems (QEMS), are very precious tools for studying mesoscopic properties. Please read articles [BLE 04, NAI 06] and [O'CO 10] on this subject.

One of the main lessons to take away from this chapter is the impact surface states have on intrinsic conduction phenomena and on near field forces occurring when the separation distance between two systems becomes small.

5

Conclusion and Application Prospects: from Fundamental Physics to Applied Physics

Having focused on the main theoretical and experimental research conducted on scaling effects and transduction methods used in nanosystems, we will conclude this book by describing a number of examples of applications for nanoelectromechanical systems (NEMS), which will be completed in the short or long term.

As has been seen, scaling effects give NEMS interesting properties for different applications. Let us refer back to the table first presented in Chapter 1 summarizing the electromechanical properties of NEMS (Table 5.1).

Parameters	Law		**Typical values**
Mass	$m \propto wlt$	α^3	1 pg–10 fg
Stiffness	$k \propto {Ewt^3}/{l^3}$	α	1 N/m–10^2 N/m
Frequency	$f \propto \sqrt{E/\rho}\, t/l^2$	α^{-1}	10 MHz–1 GHz
Dissipated mechanical energy	$P_{th} \propto {2k_BTQ}/{\pi f}$	α	100 aW–10 fW

Mechanical time constant	$\tau_m \propto {}^{Q}/_{2\pi f}$	α	0.1 µs–10 µs
Thermal time constant	$\tau_{th} \propto {}^{c\rho l^2}/_{K}$	α^2	0.1ns – 100ns
Noise amplitude	$\sqrt{{}^{2k_BTQ}/_{fk}}$	1	1 fm–100 fm
Noise in force	$\sqrt{{}^{2k_BTk}/_{fQ}}$	α	10 fN–1 pN
Mass detection limit	$\delta m = 2m\,{}^{\delta f}/_{f_0}$	α^3	(ag – yg)

Table 5.1. *Associated scaling laws when a reduction coefficient α is applied to its length, width and thickness: $l' = \alpha l$, $w' = \alpha w$, $t' = \alpha t$ (Figure 1.12). E, ρ, c and κ are the Young's modulus, volume mass, thermal capacitance and thermal capacitance of the nanowire; k_B and T are the Boltzmann constant and temperature, respectively*

Table 5.1 can be summarized as follows: NEMS constitute, in particular, mechanical systems that are high frequency, fast and extremely sensitive in detecting minuscule forces (pN) or very low masses (fg). A number of examples of applications will be described in detail below.

First, let us discuss inertial sensors manufactured by integrating silicon nanowires. The latter form the core of the sensor and are used to measure the axial constraints transmitted by a proof mass. The constraint is directly proportional to acceleration, or the rotation velocity of the mass, which turns around a rotation axis (see Figure 5.1). These structures are unique as they are manufactured with two different thicknesses of silicon: the nanowires are structured with a first thin layer (~250 nm) and connected to the proof mass which is etched into a thicker layer (~10–30 µm). Proposed by the CEA-LETI, this hybrid approach known as *M&NEMS* uses the extreme sensitivity of the NEMS and maintains a significant proof mass [ROB 09, ETT 14, WAL 12] . In so doing, these sensors give performance levels that are better than or equivalent to current performance (mg over a range +/–10g), with a substantial surface gain factor of at least 2 or 3. Multi-axis accelerometers, gyroscopes and magnetometers have been manufactured using this new architecture (three-axis accelerometer presented in

Figure 5.2) and will be targeted at countless markets, including the automobile, consumer and aerospace markets.

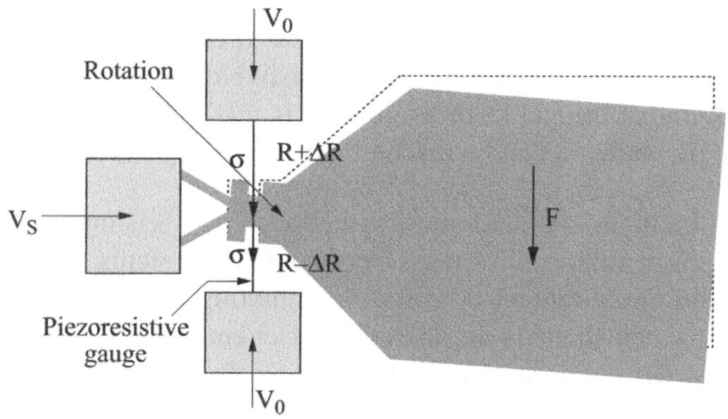

Figure 5.1. *Principle scheme of an M&NEMS-type sensor with two nanowires used in constraint gauges (etched into a Si layer with low thickness) and a proof mass etched into a thick Si layer – when acceleration occurs, the mass turns around a rotation point causing a tensile constraint in a wire and a compression constraint in the opposite wire. As the wires are piezoresistive gauges, we measure the output voltage V_s corresponding to the relative resistance variations of the gauges. This scheme shows what is known as a "mono-axis" accelerometer (in this case) but the principle has been demonstrated for three-axis accelerometers, gyroscopes and magnetometers*

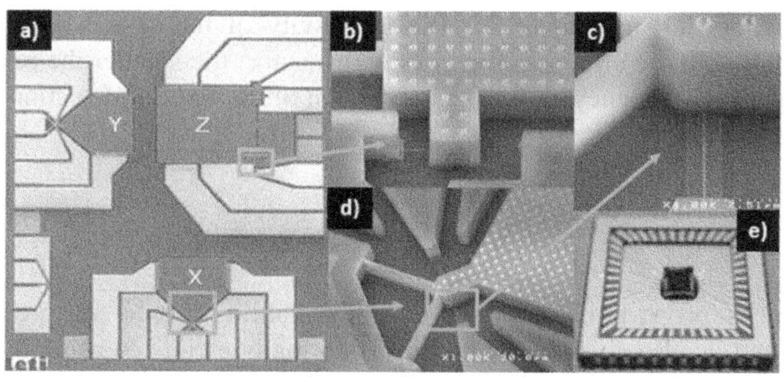

Figure 5.2. *Example of M&NEMS magnetometer sensors: a) view from above of the three sensors to measure the three axes; b) zoom of the sensitive part of the sensor axis Z; c) zoom of the sensitive part of a sensor X; d) nanowire with low thickness attached to the proof mass with high thickness and e) packaged three-axis component connected to a support to be tested*

Force can also be measured for biomedical purposes. In this case, force is cellular rather than inertial. A (biological) cell, for example, is immobilized on a set formed by a fixed pad facing a mobile nanobeam. As it contracts, the cell exerts a force on the nanobeam, thereby causing it to deform. In principle, the force is measured by piezoresistive detection. In this case in point, the NEMS must be very flexible (low mechanical stiffness) and the materials used are often polymers. Piezoresistive detection therefore happens by adding a metallic layer such as gold. It is thought that resolutions of the order of the piconewton can be reached with these objects [ARL 06].

To date, the main application envisaged for NEMS is measuring mass for gas sensors or realizing new mass spectrometry systems. As explained in Chapter 1, the measurement consists of estimating the nanoresonator's shift in frequency when there is an accretion of chemical species on its surface (see Figure 1.8). As a reminder, the shift in frequency Δf caused by the adsorption of a mass m is given by the following expression:

$$\Delta f = \frac{-m}{2m_{eff}} f_0 \qquad [5.1]$$

where f_0 is the nominal resonance frequency without loading and m_{eff} is the effective mass of the resonator.

The NEMS is put in a closed loop, such as a self-oscillating loop or a phase-locked loop (PLL), so as to follow frequency variation in real time (see section 2.3 in Chapter 2). The sensitivity of the system is inversely proportional to the effective mass of the resonator: the smaller the resonator, the larger the frequency shift. The expression of the limit of detection (LOD) defined in Table 5.1 is:

$$LOD = 2m_{eff} \, {}^{\delta f}/_{f_0} \qquad [5.2]$$

LOD is proportional to the frequency stability of the nano-oscillator $\delta f/f_0$ and its effective mass. Stability is normally estimated using the Allan variance, which was described in Chapter 2 (section 2.3, equation [2.57]). We were therefore able to state that Allan variance is inversely proportional to the signal-to-noise ratio when the noise regime is dominated by bench noises. By reducing these two parameters, the *LOD* record obtained until now corresponds to several hydrogen atoms. In this experiment, a carbon nanotube placed in low temperature (~4 K) and ultra-vacuum (10^{-9} Torr) was used [CHA 12]. More classically, the characteristics of silicon nanostructures enabled detection limits between 1 and 50 kDa (1 Da = 1.67×10^{-24} g) to be reached. This performance is sufficient for NEMS to be integrated into gas analysis systems or new generation mass spectrometers.

For gas analysis, the surface of the NEMS must be functionalized (see Figure 5.3). With such an active layer, the gas elements are retained on the NEMS surface for a duration longer than that of the measurement time. This layer acts as a sponge whose efficiency varies according to the chemical affinity of the gas in relation to the chosen layer. This efficiency is characterized by the partition coefficient K_l, which is the ratio between the concentration of chemisorbed and/or physisorbed analyte and the concentration in vapor phase. Using K_l, we can find out the additional mass adsorbed on the vibrating nanostructure, as shown by the following equation:

$$K_l = \frac{C_s}{C_v}$$
$$\delta m = C_v K_l \rho_v V_c \qquad [5.3]$$

where C_s is the concentration of analyte adsorbed into the chemical functionalization layer, C_v is the concentration of analyte in vapor phase (expressed in molar ppm), ρ_v is the density of the gas phase and V_c is the volume of the functionalization layer.

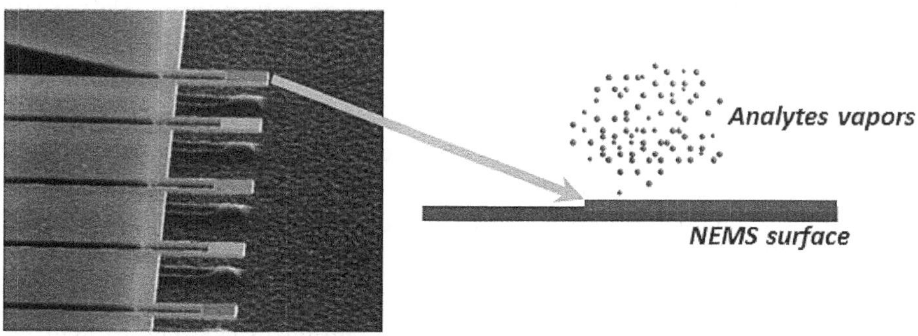

Figure 5.3. *NEMS array functionalized to adsorb the gases of interest on their surfaces. For a color version of the figure, see www.iste.co.uk/duraffourg/nems.zip*

A great deal of research has focused on the optimization of functionalization layers depending on the gases of interest by using broad categories of surface bonding[1]. Two broad types of architecture have been developed.

The first type of architecture consists of realizing electronic noses (e-nose). Each NEMS or group of NEMS in an array is functionalized for a gas group by an eigenlayer. It has been demonstrated that there is a maximum of eight groups of different couples (gas and layers) that form an orthogonal vector space enabling the odor fingerprint to be built via an adapted algorithm. Above this number, partition coefficients by gas are almost identical and the layers become redundant. It is, for example, possible to detect the presence of a tube of toothpaste in a backpack *ex post facto*. This system cannot quantify each volatile chemical compound in the toothpaste, but it recognizes its global fingerprint.

The second type of architecture is based on conventional analysis systems using the columns in gas chromatography (GC) [GRO 04, ACK 08] before the detectors. The GC column is a tube whose internal surface is functionalized by a polar or apolar chemical layer (known as the stationary phase – SP) according to the broad categories of gases to analyze. The mean velocity at which each gas passes through the column depends on its affinity with the active layer. Affinity sets the length of time the gas is imprisoned in the SP. The gases exit the column as elution peaks separated over time by their chemical affinity with the SP. The operating principle of the chromatography column is described in Figure 5.4. The NEMS placed at the end of the column act as a universal quantification system. In this architecture, the NEMS are functionalized by a generic layer so that sensitivity is almost uniform for all gases. The different functions of this architecture are shown in Figure 5.5. The gas mixture is injected into the column by an injector controlling the temperature and the pressure (or by a MEMS microinjector). The elution peaks separated over time are then measured by NEMS located at the output of the GC. Concentration is evaluated by integrating the area under the elution peaks. The system has three characteristic principles in comparison to commercial systems: (1) the size of the NEMS in relation to the countless sensors available enables packaging/a fluid channel with sections similar to that of GCs or micro-GCs. This characteristic enables the effect of dead volumes to be limited, which are a source of the dispersion of elution peaks. In other words, fluidic

1 The gases connect to the surface by pure physisorption (Van der Waals bonding, hydrogen) or by chemisorption (covalent bonding).

impedance of NEMS packaging matches the capillary impedance. Separation is therefore improved by the use of NEMS; (2) the response time of the NEMS makes it possible for measurement time to be lower than 20ms, which then enables elution peaks to be accurately sampled for several seconds so as to realize a precise integration of the area below the peak; (3) the LOD of the NEMS is extremely low. For instance, NEMS can detect BTEX[2] concentrations of 10 ppb with dynamic measurement (i.e. with a GC over 20 ms of integration time). Figure 5.6 shows a chromatogram of a BTEX gas mixture at concentrations of 1 ppm.

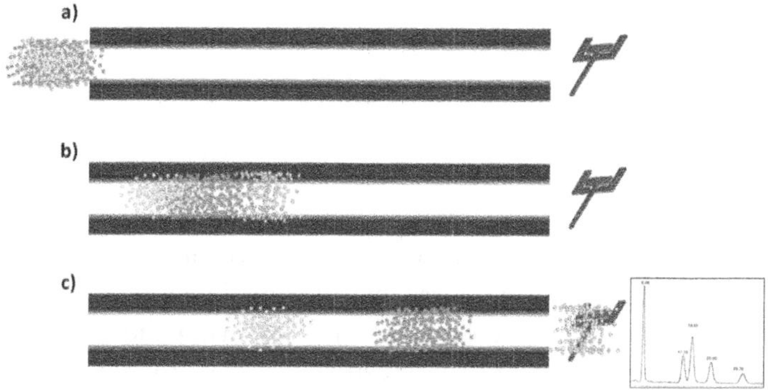

Figure 5.4. *Operating principle of a chromatography column in gas phase: a) injected gas mixture; b) progression of gases in the column; the propagation velocity of the gases differs according to the affinity of these gases with the stationary phase (black layers); c) gases exit as elution peaks (shown on the right of the NEMS) separated over time. For a color version of the figure, see www.iste.co.uk/duraffourg/nems.zip*

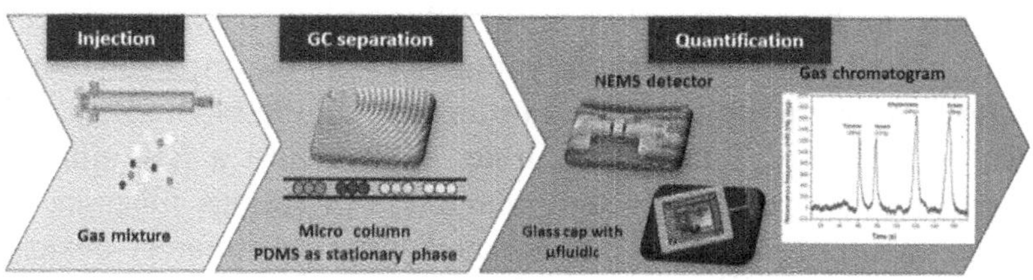

Figure 5.5. *Description of the measurement chain including an injection of the gas mixture, a chromatography column in gas phase realizing the separation (identification) and a NEMS or NEMS array realizing quantification (measurement of the concentration corresponds to the integration of the signal under the elution peak). For a color version of the figure, see www.iste.co.uk/duraffourg/nems.zip*

2 BTEX: benzene, toluene, ethylbenzene and xylene.

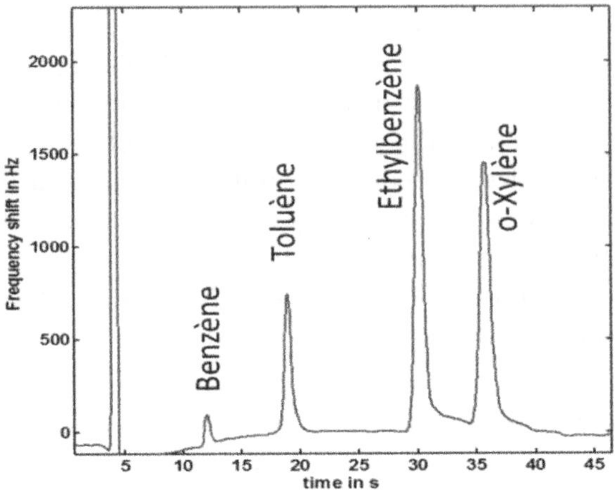

Figure 5.6. *Chromatogram of a BTEX gas mixture*

These systems are used in a large number of applications that require chemical analysis: for instance, gas monitoring in industrial environments (petrochemistry, chemical factories, etc.), measuring the quality of indoor and outdoor air, agro-foods, domestic security (detecting explosives, biochemical agents, etc.) and biomedical domains (diagnostics). Regardless of the application, the standard measurement tool is the mass spectrometer coupled with a GC. The GC/NEMS system presented in this section (see Figure 5.5) is capable of achieving the performances required to eventually replace the standard analysis system. The total volume of this analysis system ranges from several m^3 to 1 L (see Figure 5.7), and as such becomes portable and can be distributed on sites. When this tool eventually becomes available, it will profoundly modify practices in the medical domain by considerably accelerating research into identifying the biomarkers in breath or given out by body fluids (skin, urine, etc.). In just a few minutes, this device will be able to identify the biomarkers of serious illnesses such as lung cancer, neurodegenerative diseases and chronic respiratory disorders.

Distributing this technology in hospitals, clinics, analysis laboratories and surgical processes will pave the way for a revolutionary approach to diagnosis and non-invasive medical screening. Taking this first step will finally promote the appearance of point-of-care (PoC)-type practice while the democratization of technologies due to miniaturization will make the tools accessible to the greatest number of people for home use.

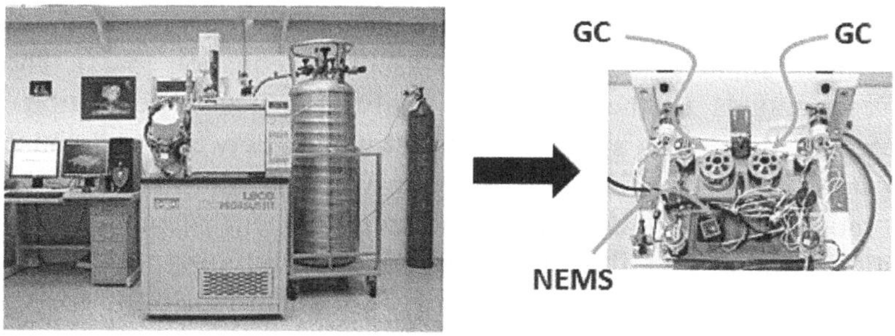

Figure 5.7. *Size and complexity of a current breath analysis system: 2D chromatography of a mass spectrometer (GCxGC-MS) compared to the GCxGC-NEMS system*

NEMS also enable a new paradigm in the shape of a new mass spectrometer to identify large biomolecules. Mass spectrometry is a universal analysis technique used practically across all scientific domains (physical chemistry, materials science, biology, etc.). The operating principle of a mass spectrometer is based on the separation per electromagnetic field of previously ionized particles according to their mass to charge ratio *m/z* (see Figure 5.8). To do this, a mass spectrometer is composed of a source of ions, an analyzer and a charge counter. A detailed description of the architecture of a mass spectrometer is beyond the scope of this chapter. A very large number of ionization techniques exist, which are chosen depending on which molecules are to be analyzed, and there are no fewer separation systems. For more details, refer to [DOM 06, BAN 07].

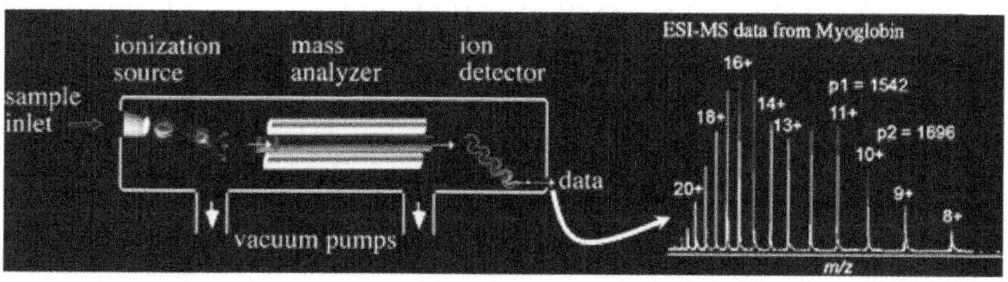

Figure 5.8. *Principle scheme of a mass spectrometer and example of a mass spectrum (source: Scripps Center for Metabolomics). For a color version of the figure, see www.iste.co.uk/duraffourg/nems.zip*

The performance of a mass spectrometer is characterized by its sensitivity, its mass range, its resolving power and its precision. Sensitivity corresponds to the minimum number of molecules of interest that need to be injected into the system to be able to conduct the analysis. In the case of NEMS, this parameter corresponds to its LOD (equation [5.2]). The mass range corresponds to the maximum analyzable mass. Resolution power characterizes the device's capacity to distinguish between two consecutive peaks. It is defined by the inverse of the width at mid-height to the mass range ratio, $m/\Delta m$. Precision is the measurement error in relation to theoretical value. For illustration purposes, orbitrap-type spectrometers [MAK 00, HU 05] have resolution powers that can reach 105 to 106 for a mass range of 5 kDa. This means that this instrument is capable of suggesting the difference between two masses separated by a fraction of a Da. Their sensitivity is of the order of 108 particles. More generally, mass spectrometers have resolution powers varying from 104 to 105 for mass ranges capable of reaching several thousand Da.

With this performance, mass spectrometers can identify small differences in mass between isotopes, analyze the composition of carbonated chemical or complex mineral elements, analyze DNA strands, and even peptides (see Figure 5.9). More recently, they have been used in proteomics (the study of proteins)[3] and in virology in wide mass ranges between 1 and 100 kDa. Research groups have expanded the mass range to several MDa to identify complex proteins or intact viruses [HEC 08, UET 08]. Nevertheless, resolution and sensitivity are greatly damaged (see Figure 5.9).

In contrast, NEMS has become competitive in this measurement range. Resolution and sensitivity are effectively fixed by *LOD*. As this mass *LOD* is typically several dozen kDa [MIL 10, YAN 06], it is possible to identify biomolecules as soon as mass is higher than this low limit (a low limit is usually considered to be 3 *LOD*). The NEMS is therefore very sensitive as it is capable of conducting a measurement biomolecule by biomolecule. Moreover, the resolving power expressed as m/LOD is just as high as the mass range considered is wide. This tendency is the exact opposite of conventional mass spectrometers, which see their resolving power slowly dwindle as the mass range increases (as shown in Figure 5.9). NEMS can, therefore, measure masses that are too high for mass spectrometers but too low for other possible techniques (such as optical techniques). Viruses and proteins or intact protein complexes can be measured. Some of these biomolecules are markers of cancers or infectious diseases, which are

3 Proteins are composed of peptides. The typical mass of the latter is 1 kDa whereas proteins normally range between 20 kDa and 1 MDa.

difficult to recognize using a classic proteomic approach. Sensitivity to single molecules also speeds up analysis. The detector does not require biological matter to be amplified with long (e.g. cultivation) and potentially costly methods.

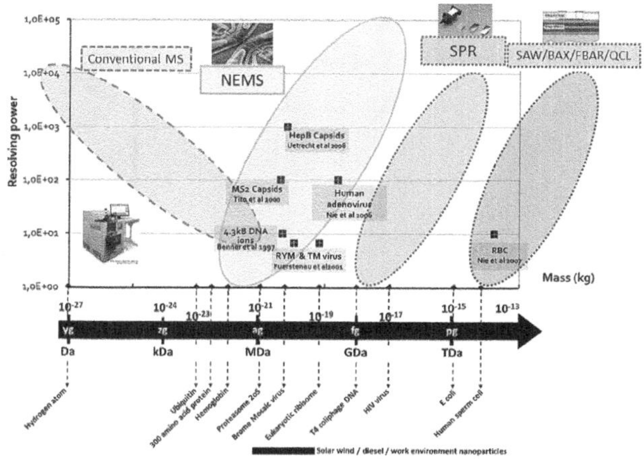

Figure 5.9. *Comparison of different mass measurement systems: comparison of their resolution power according to the mass of molecules of interest (expressed in kg, g and then Da). For a color version of the figure, see www.iste.co.uk/duraffourg/nems.zip*

For illustration purposes, we will describe the first mass spectrometry experiment conducted on IgM proteins (human antibodies) [HAN 12]. The measurement system consists of sending a protein spray to a vibrating NEMS. The NEMS is put under vacuum and the surroundings are at the low temperature of 80 K (see Figure 5.10). Its resonance frequency is monitored in real time by a PLL. Each biomolecule landing on the surface of the NEMS causes a frequency shift, a characteristic that can be quantified by equation [5.1], as shown in Figure 5.11. The latter demonstrates the NEMS' capacity to detect a single molecule. The NEMS is actually put in resonance with two modes so as to improve the measurement precision by localizing the mass. Indeed, the use of two eigenmodes results in two equations with two unknowns: the mass and the position of the mass along the nanobeam. There is a strong correlation between the frequency shifts of mode 1 and the frequency shifts of mode 2. Following the real-time treatment of the signal, the spectrum is constructed molecule by molecule (see Figure 5.12). It can be stated that the IgMs sent are actually IgM isoforms (tetramers, pentamers, etc.). This first experiment is extremely promising and demonstrates the potential of NEMS in the domain of mass spectrometry.

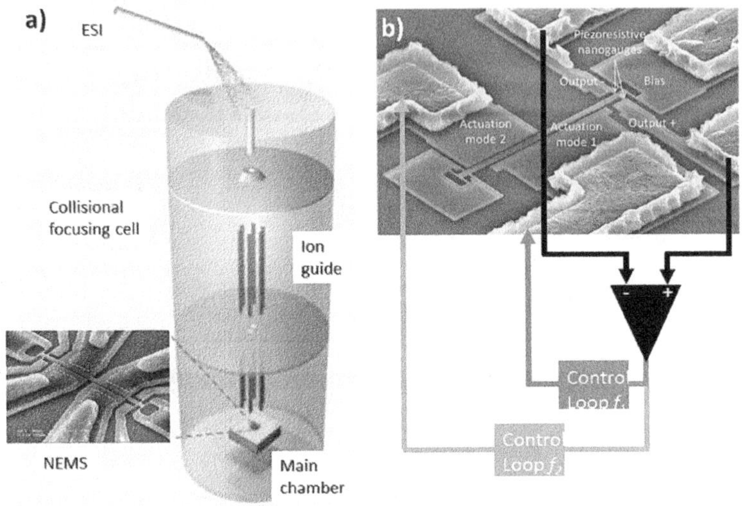

Figure 5.10. *NEMS-based spectrometry according to [HAN 12]: a) mass spectrometry system formed of an ionization source of biomolecules by spray (ESI), a hexapole ion guide and a cryogenic cane on which the NEMS is fixed – the NEMS is under vacuum and at low temperature; b) principle scheme of closed loops (PLL) that can actuate and read out signals at two frequencies corresponding to the two resonance modes of a pinned-pinned nanobeam. For a color version of the figure, see www.iste.co.uk/duraffourg/nems.zip*

The IgM spectrum was realized on a very low fraction of captured biomolecules. The surface of the NEMS is extremely small (5×0.3 μm^2) compared to the section of the spray (5×5 mm^2), meaning capture yield is 6×10^{-8}. It is therefore essential to increase the latter by only a few percentages so as to reach reasonable analysis speeds (i.e. at least equal to the speeds of current analysis). Two future approaches can therefore be predicted: the first approach consists of improving the focalization of the molecule beam and the second approach consists of increasing the capture cross section of NEMS. To do so, NEMS must be manufactured with very large-scale integration (VLSI) so that they can be put in the network (see Figures 5.13 and 5.14).

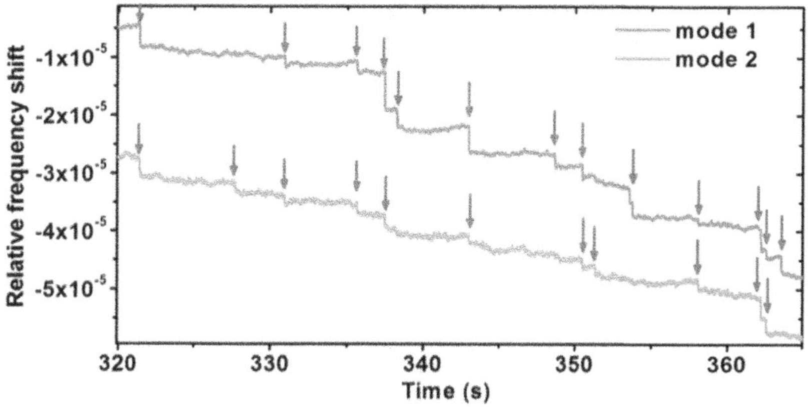

Figure 5.11. *Example of the frequency shifts observed when the particles are on the surface of the NEMS: the top curve corresponds to the first resonance mode and the bottom curve corresponds to the second resonance mode. The frequencies tend to decrease continually because the masses accumulate on the NEMS without desorbing. Desorption can be realized by heating once the measurement is concluded. For a color version of the figure, see www.iste.co.uk/duraffourg/nems.zip*

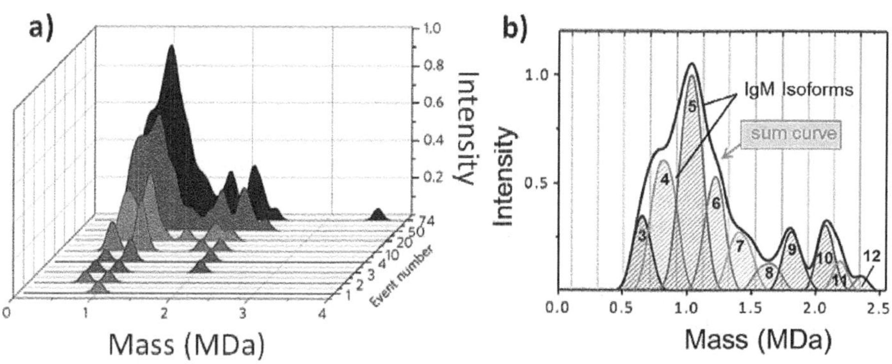

Figure 5.12. *Mass spectrum obtained on isoforms: a) events accumulating over time (real-time analysis) and b) accumulated spectrum and different contributions of each isoform at the final spectrum. For a color version of the figure, see www.iste.co.uk/duraffourg/nems.zip*

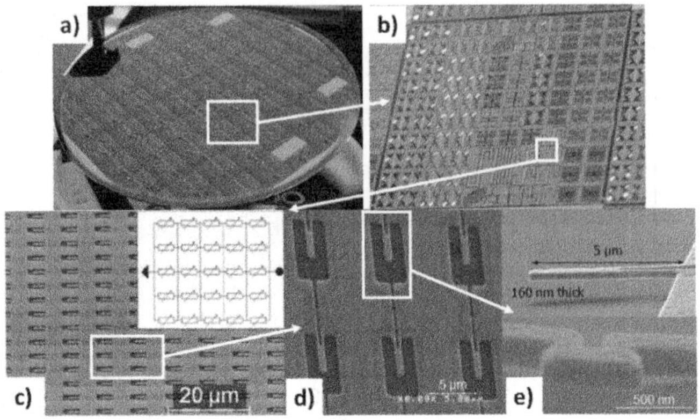

Figure 5.13. *Example of a collectively addressed NEMS network: a) 200 mm wafer comprising 3.5 million NEMS; b) picture of a 2 cm² field; c) zoom of the field on a network of 20,000 NEMS; d) and e) zoom of several elements, then of a nanocantilever*

The VLSI approach enables very dense NEMS arrays to be manufactured, as shown in Figure 5.13. They can be collectively addressed by interconnecting all the NEMS. The NEMS are therefore all actuated by the same excitation signal when the output signal corresponds to the sum of signals from the NEMS. In this configuration, an array of N NEMS provides an SNR enhancement of a factor $\sqrt{N}$. Collective addressing has been successfully used in the context of gas measurement [BAR 12, LEE 08]. The network actually has a better concentration *LOD* over the same measurement time or can reach the same *LOD* with a shorter measurement time. This improvement has been demonstrated in this publication [BAR 12]. In mass spectrometry, NEMS are sensitive to discrete events, unlike gas measurement which relies on a continuous process. Indeed, measuring discrete events (potentially affecting a single NEMS at a time) using a collectively addressed array is not foreseeable. Each nanostructure in the matrix must therefore be individually addressed. In this approach, the NEMS network is comparable to an imager composed of pixels (each of the pixels corresponding to a NEMS). One way, among others, of putting this configuration in place is to slightly vary the lengths of the suspended nanostructures. Each NEMS can thereby be identified by its resonance frequency, which varies quadratically with its length. Arrays are therefore manufactured with several levels of interconnections (see Figure 5.14(c)). This addressing, known as frequency addressing, has been demonstrated by actuating and reading the 20 NEMS in a matrix with one phase-locked loop, locking the 20 frequencies [SAG 13] in turn, as shown in Figure 5.15.

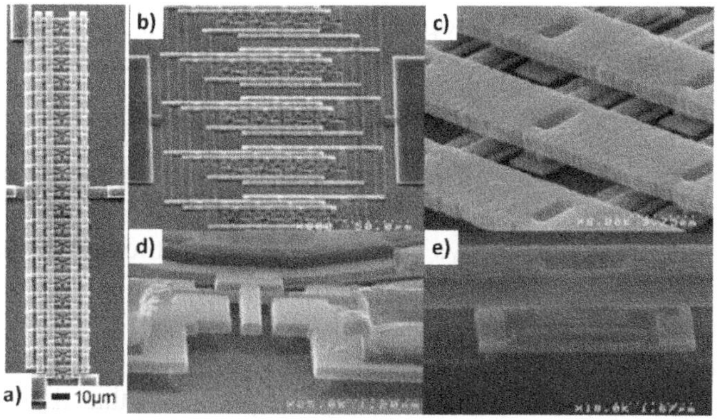

Figure 5.14. *Examples of individually addressed NEMS arrays: a) column of 20 nanocantilevers; b) another 5 × 4 NEMS matrix configuration (view from a scanning electron microscope, in color); c) zoom of a part of the metallic interconnection, enabling the addressing lines to cross; d) zoom of a cantilever; e) close-up of a metallic via realizing the connection between two metallic levels. For a color version of the figure, see www.iste.co.uk/duraffourg/nems.zip*

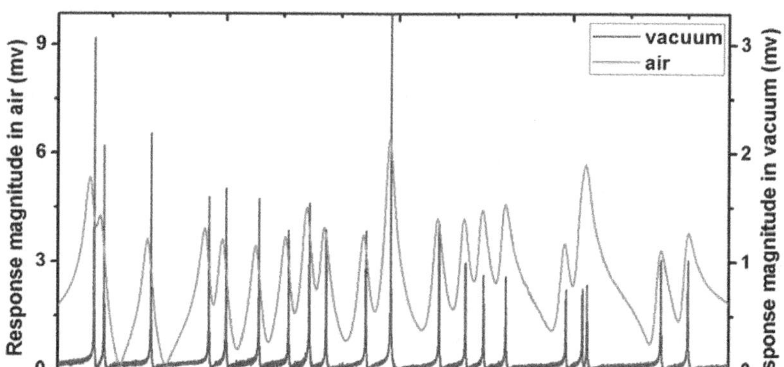

Figure 5.15. *Frequency addressing of an array of 20 NEMS [SAG 13]: the NEMS are actuated one by one by scanning the excitation frequency on the frequency range corresponding to the mechanical resonances of the nanocantilevers. The bottom curve represents the frequency responses of the NEMS under vacuum and the top curve represents the frequency responses of the same NEMS under air. As the quality factor is lower, the resonance peaks tend to recover. For a color version of the figure, see www.iste.co.uk/duraffourg/nems.zip*

In conclusion, NEMS are open to a wide field of applications: (1) the biomedical domain for the identification of biomarkers and therapeutic treatment supervision, (2) the industrial domain (petrochemistry, agro-food chemistry, etc.), (3) the domain of personal security, (4) the domain of national security and even (5) the domain of monitoring indoor and outdoor air quality. All these cases enable the chemical elements of interest to be identified with great sensitivity. The reduced size of the NEMS has changed the sensor's integration approach in analysis instruments. Researchers have adapted the size of the fluid and electrical interfaces to the size of the NEMS or NEMS array. In addition to intrinsic performance, future analysis devices will be very compact and portable. It can be predicted that their use will be widespread across civil, medical and military societies. The small number of examples presented in this book is not an exhaustive list of domains covered by NEMS. Although briefly mentioned, nanorelays and inertial nanosensors, for example, could have been more broadly developed.

We still have a long way to go before instruments containing NEMS sensors will be commercialized. Efforts are currently focusing on improving the electrical and fluid interfacing of NEMS. Co-integration, discussed in Chapter 3, is the subject of intensive research. In Chapter 5, we learned that NEMS must be put in an array to improve the performance, more specifically facing the capture cross section (compared, for example, with the section of a chromatography column or with the surface of a biomolecule spray). Researchers have henceforth been trying to develop NEMS that are co-integrated with the electronics such that each NEMS pixel in an array is controlled by an electronic circuit (see Figure 5.16). Current manufacturing technologies enable us to contemplate three-dimensional (3D)-type integrations. The electrical transduction methods described in Chapters 2 and 4, which seem to be the most easily integrable with VLSI manufacturing processes, are piezoresistive detection or detection by field effect without junction and purely electrostatic excitation.

Finally, optomechanical transduction, which consists of actuating by optical force and detecting the vibrations of the NEMS with one of the techniques shown in this chapter, enables us to free ourselves from the electronics. NEMS array would therefore be addressed by a photonic circuit, the benefit of which is its extremely wide bandwidths inherent to the optics, thus enabling it to address a large number of NEMS simultaneously. The latter approach remains, however, prospective.

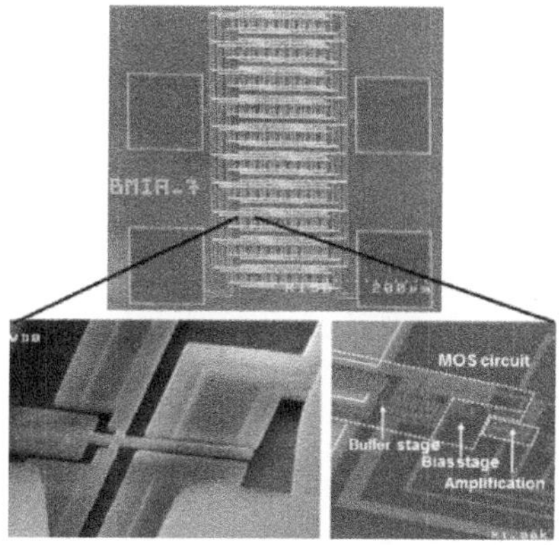

Figure 5.16. *Co-integrated NEMS array (VLSI technology) with its electronics*

Appendix

A1.1. "Bottom-up" and "top-down" manufacturing processes for nanowires

The difference between the two processes for manufacturing suspended nanowires is as follows: the "bottom-up" process can be defined as additive technology, whereas the "top-down" process can be seen as elimination technology.

A1.1.1. *"Bottom-up" manufacturing*

More specifically, manufacturing is realized by growth processes using a precursor. These processes are based on vapor–liquid–solid (VLS)-type growth as shown in the figures below. The nanowire is built from a pattern of metallic pads (usually gold). This metal is used as a precursor that chemically reacts with a gas. In the case of silicon nanowires, this gas is often silane. Many of the various variants are detailed in [SCH 09].

The VLS technique enables us to obtain almost macroscopic wires (with diameters of 100 μm) as well as nanowires with a diameter of several nanometers. These wires can be placed vertically or horizontally in relation to the support depending on the position of the metallic pads.

A scanning electron microscopy (SEM) photograph of a nanowire forest and a transmission electron microscopy (TEM) zoom of a nanowire obtained by the VLS process are shown below (Figure A.1).

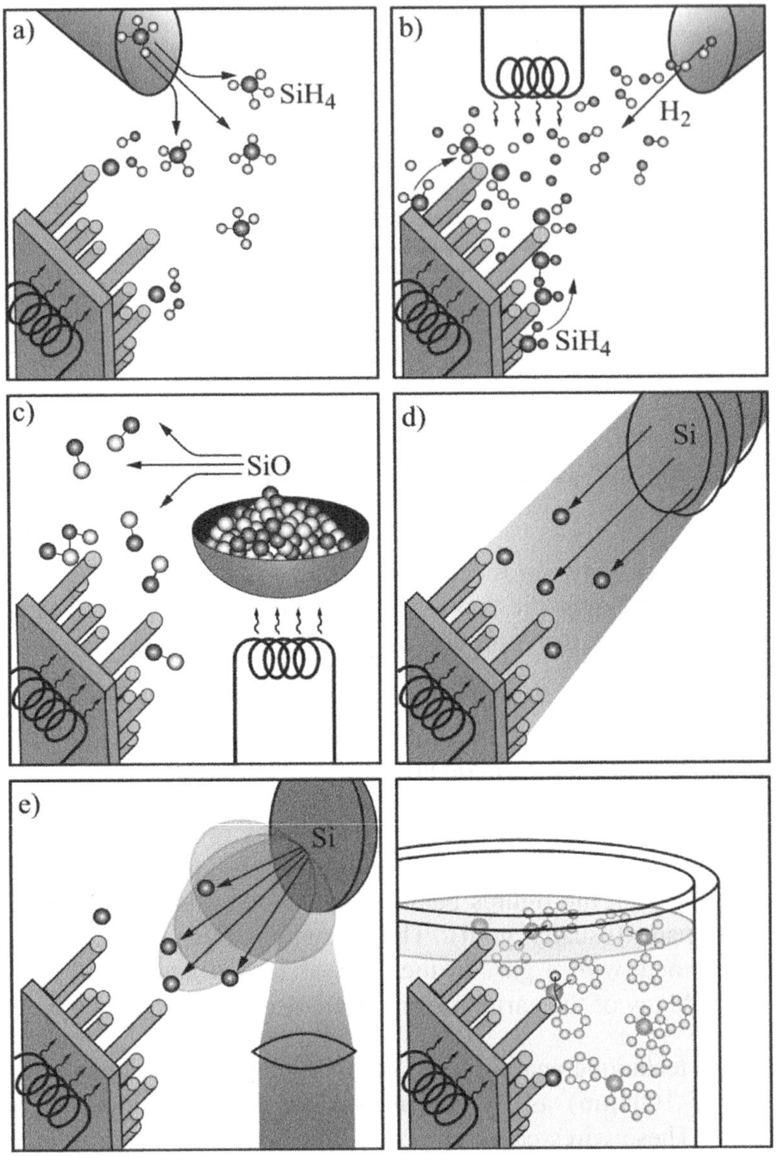

Figure A.1. *Nanowire growth by different methods (from [SCH 09]): a) chemical vapor deposition (CVD); b) annealing under reactive atmosphere; c) evaporation of* SiO_2*; d) molecular beam epitaxy (MBE) by molecular jet; e) ablation laser; f) growth in reactive solution*

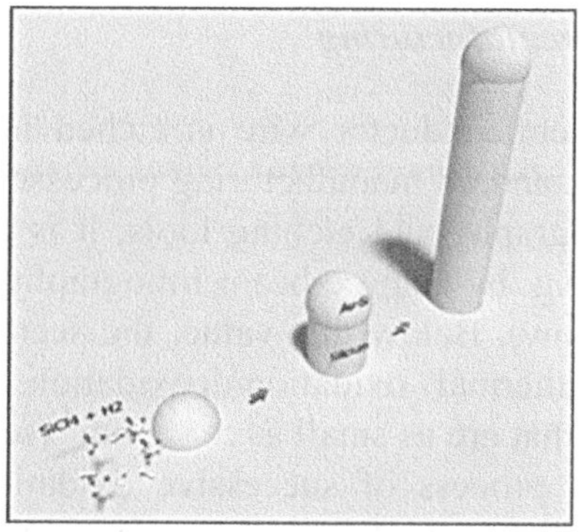

Figure A.2. *Nanowire growth by the VLS technique*

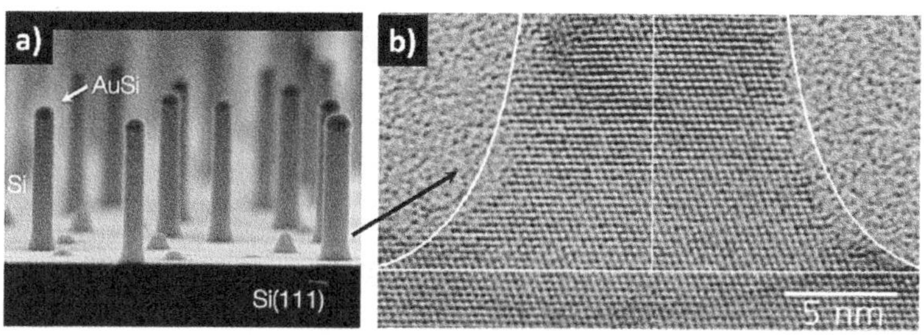

Figure A.3. *Example of nanowires obtained by the VLS technique; a) source nanowires (picture from Georgiatech Institute – Filler Group); b) cross-section in the height of the nanowire [SCH 09] (picture from the Max Planck Institute)*

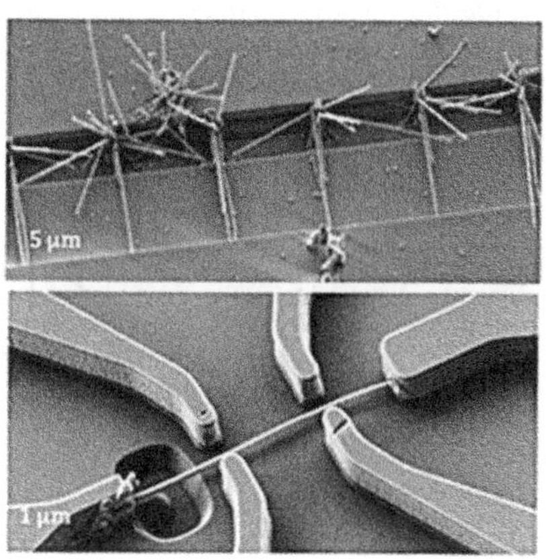

Figure A.4. *Example of horizontal nanowires obtained by VLS technique [FER 13]*

A1.1.2. *"Top-down" manufacturing*

In this case, the semiconductor wire is etched into the material using processes inherent to standard manufacturing processes for microelectronics. By using classic lithography and etching tools, it is easy enough to obtain diameters of 50 nm (e.g. by using e-beam lithography – writing by electron beam) (see figure below). Below this value, the section can be reduced by successive steps of thermal oxidation/deoxidation. Round sections can therefore be obtained that are as small as 5×5 nm^2 (see figure below). It has been shown that this process of successive oxidations was self-limiting, thereby enabling the diameter of the wire to be very accurately controlled [ERN 09].

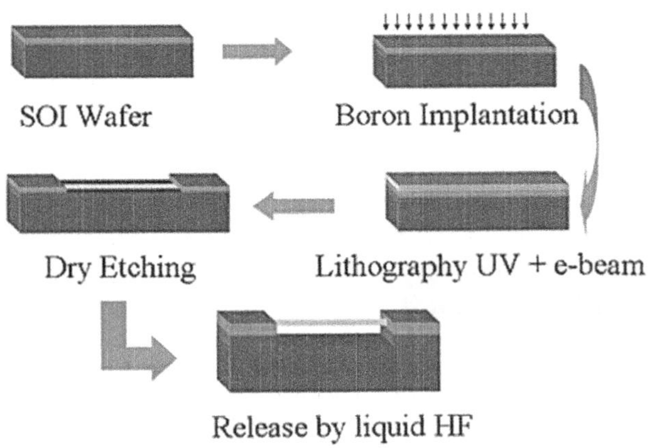

Figure A.5. *Example of a top-down CMOS process to obtain a nanowire [AYA 07]. It is possible to implant the surface in order to control the level of doping. A step of classic or even hybrid lithography superimposing an ultraviolet (UV) lithography and e-beam writing is used. The wire is then etched and the oxide layer placed below is chemically etched (HF: hydrofluoric acid). For a color version of the figure, see www.iste.co.uk/duraffourg/nems.zip*

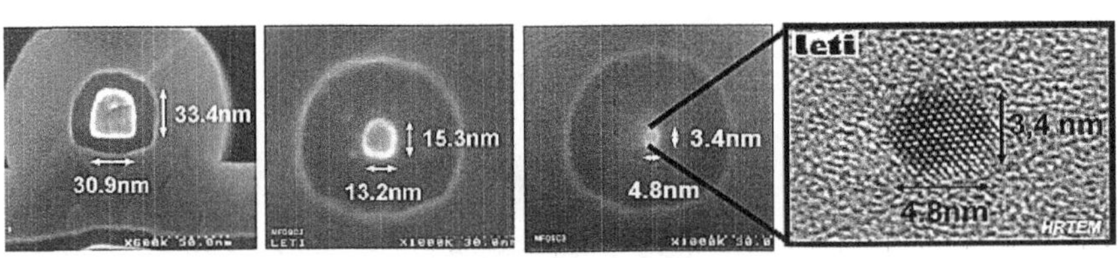

Figure A.6. *Oxidation/deoxidation process for reducing the section of the silicon wire [ERN 09]: the starting nanowire has a square section of 50×50 nm^2. After the successive steps, the section fills out and has a diameter of 4 nm*

A1.2. Casimir force in detail

To calculate the Casimir force, the sum of the pressures inside and outside the cavity must be realized:

$$\langle \hat{P} \rangle = \sum_m \iint \frac{dk_x dk_y}{4\pi^2} \int \frac{dk_z}{2\pi} \cos^2\left(\theta_m(k_x, k_y, k_z)\right) \hbar\omega_m(k_x, k_y, k_z)\left(1 - A_m(k_x, k_y, k_z)\right)$$

where (k_x, k_y, k_z) is the wave vector of the incident electromagnetic wave. The symbol m represents the polarizations s and p. $\theta_m(k_x, k_y, k_z)$ is the incidence angle of the field on the slabs. $\omega_m(k_x, k_y, k_z)$ is the pulsation which, of course, depends on the wave vector through the dispersion relation. $A_m(k_x, k_y, k_z)$ is the Airy function of the cavity that depends on the reflection coefficients $r_{mslab}(\omega, k_z)$ of each blade:

$$A_m(k_x, k_y, k_z) = \frac{1 - \left|r_{mslab}(k_x, k_y, k_z)\right|^2}{\left|1 - r_{mslab}(k_x, k_y, k_z)^2 \exp(2ik_z d)\right|^2}$$

where $r_{mslab}(\omega, k_z)$ is a function of the thickness of the slabs w and the dielectric permittivity of the material composing them. This function is simply the intensity reflection coefficient of the slab. The slab reflection itself depends on the reflection coefficients of each air/silicon and silicon/air diopter. The exponential contains the phase term $2k_z d$ representing a round trip in the cavity.

The Airy function shows the fact that the cavity selects the resonance eigenmodes proportional to $c/2d$. There are therefore fewer optical modes internally than externally. External pressure is therefore higher than internal pressure, which tends to bring the two planes closer together.

It is easier to work in the complex plane to realize integration in order to avoid problems of divergence at the poles. The expression of Airy function is:

$$(1 - A_m(\omega,k_z)) \equiv f_m(\omega,k_z) + f_m^*(\omega,k_z)$$

$$f_m(\omega,k_z) = \frac{r_{mslab}^2(\omega,k_z)\exp(2ik_z d)}{1 - r_{mslab}^2(\omega,k_z)\exp(2ik_z d)}$$

It is also preferable to work with variables without dimensions:

$$k_z = i\kappa \qquad K = \kappa d$$

$$\omega = i\xi \qquad \Omega = \frac{\xi d}{c}$$

$$x = \frac{w}{d}\sqrt{K^2 + \Omega^2(\varepsilon - 1)}$$

$$\varepsilon = \varepsilon(\omega) = \varepsilon(i\frac{c\Omega}{d})$$

K is the phase shift caused by a round trip in the cavity projected onto the propagation axis z and normalized. Ω is the analog of normalized pulsation. The expression of x corresponds to the phase term in each normalized slab at d.

Finally, the general expression of the correction factor η according to the thickness of the slabs and the material used is as follows:

$$\eta = \frac{120}{\pi^4}\sum_m \int_0^{+\infty} dK K^3 \int_0^K \frac{d\Omega}{K} f(r_m^2, K, x)$$

$$f(r_m^2, K, x) = \frac{r_m^2(1 - e^{-2x})^2 e^{-2K}}{(1 - r_m^2 e^{-2x})^2 - r_m^2(1 - e^{-2x})^2 e^{-2K}}$$

$$r_p = \frac{\sqrt{K^2 + \Omega^2(\varepsilon - 1)} - \varepsilon K}{\sqrt{K^2 + \Omega^2(\varepsilon - 1)} + \varepsilon K} \qquad r_s = \frac{\sqrt{K^2 + \Omega^2(\varepsilon - 1)} - K}{\sqrt{K^2 + \Omega^2(\varepsilon - 1)} + K}$$

where $R_{p,s}$ are the reflection coefficients of the diopters in polarization p and s. The correction factor can easily be integrated at this stage by a digital method.

It is also possible to explain the difference between the intrinsic case and the doped case through the surface phonon–polariton coupling occurring at high frequencies (UV transitions) and the surface plasmon–polariton coupling occurring at low frequencies (free carriers brought by doping) as shown in the figures below. As doping only acts on the number of free carriers available in the silicon, its effect is only visible in surface plasmon–polariton coupling. Coupling energy is therefore much greater than doping is high. In both cases, we can tell the difference between the linking modes corresponding to the attractive part of the force and the antibonding and bonding modes corresponding to repulsive contribution.

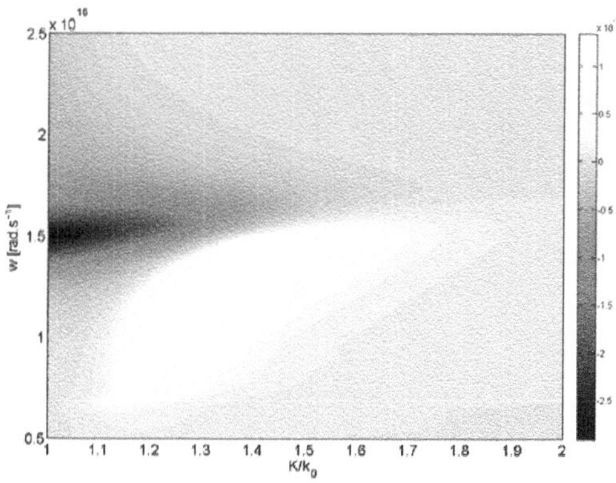

Figure A.7. *Energy density in the plane (k,w) for thin silicon slabs (100 nm) and a gap of 100 nm, surface phonon–polariton contribution (at high frequency). Calculation inspired from Greffet et al. [HEN 04]. For a color version of the figure, see www.iste.co.uk/duraffourg/nems.zip*

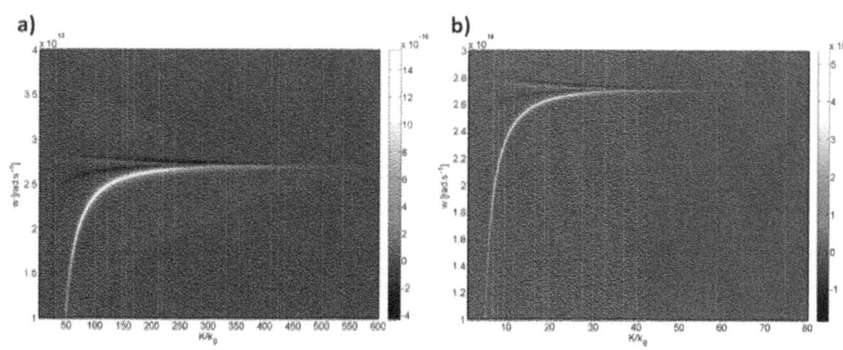

Figure A.8. *Energy density in the plane (k,w) for thin silicon slabs (100 nm) and a gap of 100 nm – surface plasmon–polariton contribution (low frequency): a) low doping P at 10^{18} cm^{-3} and b) high doping P at 10^{20} cm^{-3}. Calculation inspired from Greffet et al. [HEN 04]. For a color version of the figure, see www.iste.co.uk/duraffourg/nems.zip*

Bibliography

[ABE 06] ABELÉ N. *et al.*, "1T MEMS memory based on suspended gate MOSFET", *Proceedings of the IEEE IEDM*, pp. 1–4, 2006.

[ACK 08] ACKERMANN B.L., BERNA M.J., ECKSTEIN J.A. *et al.*, "Current applications of liquid chromatography/mass spectrometry in pharmaceutical discovery after a decade of innovation", *Annual Review of Analytical Chemistry*, vol. 1, pp. 357–396, 2008.

[AKA 09] AKARVARDAR K., WONG H.-S.P., "Analog nanoelectromechanical relay with tunable transconductance", *Electron Device Letters*, vol. 30, pp. 1143–1145, 2009.

[AKS 03] AKSYUK V.A. *et al.*, "Beam-steering micromirrors for large optical", *Journal of Lightwave Technology*, vol. 21, pp. 634–642, 2003.

[AND 06] ANDREUCCI P. *et al.*, "Impact of Casimir force on nano accelerometers", *Proceedings of the IEEE Sensors Conference*, pp. 1057–1060, 2006.

[AND 07] ANDREUCCI P. *et al.*, "Impact of Casimir force on nano accelerometers modeling", *Proceedings of the Transducers Conference*, vol. 33, pp. 1681–1684, 2007.

[ANE 08] ANETSBERGER G., RIVIÈRE R., SCHLIESSER A. *et al.*, "Ultralow-dissipation optomechanical resonators on a chip", *Nature Photonics*, vol. 2, pp. 627–633, 2008.

[ANE 09] ANETSBERGER G. *et al.*, "Near-field cavity optomechanics with nanomechanical oscillators", *Nature Physics*, vol. 5, pp. 909–914, 2009.

[ARC 08] ARCAMONE J. *et al.*, "Full-wafer fabrication by nanostencil lithography of micro/nanomechanical mass sensors monolithically integrated with CMOS", *Nanotechnology*, vol. 19, 305302, 2008.

[ARC 10] ARCAMONE J., COLINET E., NIEL A. *et al.*, "Efficient capacitive transduction of high-frequency micromechanical resonators by intrinsic cancellation of parasitic feedthrough capacitances", *Applied Physics Letters*, vol. 97, 043505, 2010.

[ARC 11] ARCAMONE J. *et al.*, "VLSI silicon multi-gas analyzer coupling gas chromatography and NEMS detectors", *Proceedings of the IEEE IEDM*, pp. 669–672, 2011.

[ARC 12] ARCAMONE J. *et al.*, "VLSI platform for the monolithic integration of single-crystal Si NEMS capacitive resonators with low-cost CMOS", *Proceedings of the IEEE IEDM*, pp. 359–362, 2012.

[ARC 14] ARCAMONE J. *et al.*, "VHF NEMS-CMOS piezoresistive resonators for advanced sensing applications", *Nanotechnology*, vol. 25, 435501, 2014.

[ARD 12] ARDITOA R., FRANGIA A., CORIGLIANOA A. *et al.*, "The effect of nano-scale interaction forces on the premature pull-in of real-life micro-electro-mechanical systems", *Microelectronics Reliability*, vol. 52, pp. 271–281, 2012.

[ARL 06] ARLETT J.L., MALONEY J.R., GUDLEWSKI B. *et al.*, "Self-sensing micro and nanocantilevers with Attonewton-scale force resolution", *Nano Letters*, vol. 6, pp. 1000–1006, 2006.

[ARN 11] ARNDT G. *et al.*, "A design methodology for fully integrated MEMS and NEMS Pierce oscillators", *Sensors and Actuators A: Physical*, vol. 172, pp. 293–300, 2011.

[ARN 12] ARNDT G. *et al.*, "Towards ultra-dense arrays of VHF NEMS with FDSOI-CMOS active pixels for sensing applications", *Proceedings of the IEEE ISSCC Conference*, vol. 54, pp. 320–322, 2012.

[ASP 13] ASPELMEYER M., KIPPENBERG T.J., MARQUARDT F. (eds), *Cavity Optomechanics*, arXiv 1303.0733, pp. 1–65, 2013.

[AYA 07] AYARI A. *et al.*, "Self-oscillations in field emission nanowire mechanical resonators: a nanometric dc – ac conversion", *Nano Letters*, vol. 7, pp. 2252–2257, 2007.

[AYM 83] AYMERICH-HUMER X., SERRA-MESTRES F., MILLAN J., "A generalized approximation of the Fermi-Dirac integrals", *Journal of Applied Physics*, vol. 54, pp. 2850–2851, 1983.

[BAL 05] BALTES H. *et al.*, *CMOS-MEMS (Advanced Micro & Nanosystems)*, Wiley-VCH Verlag, Berlin, vol. 2, 2005.

[BAN 00] BANNON F.D., MEMBER S., CLARK J.R. *et al.*, "High-Q HF microelectromechanical filters", *IEEE Journal of Solid-State Circuits*, vol. 35, pp. 512–526, 2000.

[BAN 07] BANTSCHEFF M., SCHIRLE M., SWEETMAN G. *et al.*, "Quantitative mass spectrometry in proteomics: a critical review", *Analytical and Bioanalytical Chemistry*, vol. 389, pp. 1017–1031, 2007.

[BAO 02] BAO M., YANG H., YIN H. *et al.*, "Energy transfer model for squeeze-film air damping in low vacuum", *Journal of Micromechanics and Microengineering*, vol. 12, pp. 341–346, 2002.

[BAR 05] BARGATIN I., MYERS E.B., ARLETT J. *et al.*, "Sensitive detection of nanomechanical motion using piezoresistive signal downmixing", *Applied Physics Letters*, vol. 86, p. 133109, 2005.

[BAR 07] BARGATIN I., KOZINSKY I., ROUKES M.L., "Efficient electrothermal actuation of multiple modes of high-frequency nanoelectromechanical resonators", *Applied Physics Letters*, vol. 90, p. 093116, 2007.

[BAR 09] BARLIAN B.A.A., PARK W., MALLON J.R. *et al.*, "Review: semiconductor piezoresistance for microsystems", *Proceedings of the IEEE*, vol. 97, pp. 513–552, 2009.

[BAR 11] BARBOUR N., HOPKINS R., KOUREPENIS A., "Inertial MEMS Systems and Applications", Lectures of NATO Science and Technology Organization – Low-Cost Navigation Sensors and Integration Technology, vol. 116, pp. 1–18, 2011.

[BAR 12a] BARGATIN I. *et al.*, "Large-scale integration of nanoelectromechanical systems for gas sensing applications", *Nano Letters*, vol. 12, pp. 1269–1274, 2012.

[BAR 12b] BARTSCH S.T., DUPRÉ C., OLLIER E. *et al.*, "Resonant-body silicon nanowire field effect transistor without junctions", *Proceedings of the IEEE IEDM 2013*, vol. 450, pp. 225–226, 2012.

[BAR 12c] BARTSCH S.T., LOVERA A., GROGG D. *et al.*, "Nanomechanical silicon resonators with intrinsic tunable gain and sub-NW power consumption", *ACS Nano*, vol. 6, pp. 256–264, 2012.

[BAR 12d] BARTSCH S.T., RUSU A., IONESCU A.M., "Phase-locked loop based on nanoelectromechanical resonant-body field effect transistor", *Applied Physics Letters*, vol. 101, p. 153116, 2012.

[BAR 14] BARTSCH S.T., ARP M., IONESCU A.M., "Junctionless silicon nanowire resonator", *IEEE Journal of the Electron Devices Society*, vol. 2, pp. 8–15, 2014.

[BAT 07] BATRA R.C., PORFIRI M., SPINELLO D., "Effects of Casimir force on pull-in instability in micromembranes", *Europhysics Letters*, vol. 77, 20010, 2007.

[BAT 12] BATUDE P. *et al.*, "3-D sequential integration: a key enabling technology for heterogeneous co-integration of new function with CMOS", *IEEE Journal on Emerging and Selected Topics in Circuits and Systems*, vol. 2, pp. 714–722, 2012.

[BEL 05] BELL D.J., LU T.J., FLECK N. *et al.*, "MEMS actuators and sensors: observations on their performance and selection for purpose", *Journal of Micromechanics and Microengineering*, vol. 15, pp. S153–S164, 2005.

[BIL 09] BILHAUT L., Actionnement magnétique à l'échelle nanométrique, PhD thesis, Joseph-Fourier University, Grenoble, France, pp. 1–183, 2009.

[BJÖ 09] BJÖRK M.T., SCHMID H., KNOCH J. *et al.*, "Donor deactivation in silicon nanostructures", *Nature nanotechnology*, vol. 4, pp. 103–107, 2009.

[BLA 03] BLAIS A., "Algorithmes et architectures pour ordinateurs quantiques supraconducteurs", *Annales de Physique*, vol. 28, pp. 1–148, 2003.

[BLE 04] BLENCOWE M.P., "Quantum electromechanical systems", *Physics Reports*, vol. 395, pp. 159–222, 2004.

[BLE 08] BLENCOWE M.P., "SQUIDs at the limit", *Nature Physics*, vol. 4, pp. 8–9, 2008.

[BOS 00] BOSTRÖM M., SERNELIUS B.E., "Thermal effects on the Casimir force in the 0.1-5 mm range", *Physical Review Letters*, vol. 84, pp. 4757–4760, 2000.

[BRA 85] BRAGINSKY V.B., MITROFANOV V.P., PANOV V.I., *Systems with Small Dissipation*, The University of Chicago Press, Chicago and London, 1985.

[BRA 92] BRAGINSKY V.B., KHALILI F.Y., *Quantum Measurement*, Cambridge University Press, Cambridge, 1992.

[BUS 98] BUSTILLO J.M., HOWE R.T., MULLER R.S., "Surface micromachining for micro-electro mechanical systems", *Proceedings of the IEEE*, vol. 86, pp. 1552–1574, 1998.

[CAH 03] CAHILL D.G. *et al.*, "Nanoscale thermal transport", *Journal of Applied Physics*, vol. 93, pp. 793–818, 2003.

[CAL 85] CALLEN H.B., *Thermodynamics and an Introduction to Thermostatistics*, Wiley, New York, 1985.

[CAP 01] CAPASSO F., "Quantum mechanical actuation of microelectromechanical systems by the Casimir force", *Science*, vol. 291, pp. 1941–1944, 2001.

[CAR 07] CARMON T., VAHALA K., "Modal spectroscopy of optoexcited vibrations of a micron-scale on-chip resonator at greater than 1 GHz frequency", *Physical Review Letters*, vol. 98, pp. 123–901, 2007.

[CAS 48] CASIMIR H.B.G., POLDER D., "The influence of retardation on the London–Van der Waals forces", *Physical Review*, vol. 73, pp. 360–372, 1948.

[CAV 82] CAVES C.M., "Quantum limits on noise in linear amplifiers", *Physical Review D*, vol. 26, pp. 1817–1839, 1982.

[CHA 01] CHAN H.B., AKSYUK V.A., KLEIMAN R.N. *et al.*, "Nonlinear micromechanical Casimir oscillator", *Physical Review Letters*, vol. 87, pp. 211801-1–211801-4, 2001.

[CHA 12] CHASTE J. *et al.*, "A nanomechanical mass sensor with yoctogram resolution", *Nature Nanotechnology*, vol. 7, pp. 301–304, 2012.

[CHO 09] CHONG S. *et al.*, "Nanoelectromechanical (NEM) relays integrated with CMOS SRAM for improved stability and low leakage", *Proceedings of the IEEE ICCAD*, pp. 478–484, 2009.

[CLE 02a] CLELAND A.N., ALDRIDGE J.S., DRISCOLL D.C. *et al.*, "Nanomechanical displacement sensing using a quantum point contact", *Applied Physics Letters*, vol. 81, p. 1699, 2002.

[CLE 02b] CLELAND A.N., ROUKES M.L., "Noise processes in nanomechanical resonators", *Journal of Applied Physics*, vol. 92, pp. 2758–2769, 2002.

[CLE 09] CLELAND A.N., "Photons refrigerating phonons", *Nature Physics*, vol. 5, pp. 458–460, 2009.

[COH 98] COHEN-TANNOUDJI C.N., "Manipulating atoms with photons", *Reviews of Modern Physics*, vol. 70, pp. 707–719, 1998.

[COL 90] COLINGE J., "Conduction mechanisms in thin-film accumulation-mode SOI p-channel MOSFET's", *IEEE Transactions on Electron Devices*, vol. 37, pp. 718–723, 1990.

[COL 09a] COLINET E. *et al.*, "Self-oscillation conditions of a resonant nanoelectromechanical mass sensor", *Journal of Applied Physics*, vol. 105, p. 124908, 2009.

[COL 09b] COLINET E. *et al.*, "Ultra-sensitive capacitive detection based on SGMOSFET compatible with front-end CMOS process", *IEEE Journal of Solid-State Circuits*, vol. 44, pp. 1–11, 2009.

[COL 10] COLINGE J. *et al.*, "Nanowire transistors without junctions", *Nature Nanotechnology*, vol. 5, pp. 225–229, 2010.

[COU 01] COURTY J., HEIDMANN A., PINARD M., "Quantum limits of cold damping with optomechanical coupling", *The European Physical Journal D*, vol. 17, pp. 399–408, 2001.

[DIA 07] DIARRA M., NIQUET Y., DELERUE C. *et al.*, "Ionization energy of donor and acceptor impurities in semiconductor nanowires: importance of dielectric confinement", *Physical Review B*, vol. 75, 045301, 2007.

[DIN 01] DING J., WEN S., MENG Y., "Theoretical study of the sticking of a membrane strip in MEMS under the Casimir effect", *Journal of Micromechanics and Microengineering*, vol. 11, pp. 202–208, 2001.

[DOM 06] DOMON B., AEBERSOLD R., "Mass spectrometry and protein analysis", *Science*, vol. 312, pp. 212–218, 2006.

[DRE 07] DRESSELHAUS M.S. *et al.*, "New directions for low-dimensional thermoelectric materials", *Advanced Materials*, vol. 19, pp. 1043–1053, 2007.

[DUR 06] DURAFFOURG L., ANDREUCCI P., "Casimir force between doped silicon slabs", *Physics Letters A*, vol. 359, pp. 406–411, 2006.

[DUR 08a] DURAFFOURG L. *et al.*, "Compact and explicit physical model for lateral metal-oxide-semiconductor field-effect transistor with nanoelectromechanical system based resonant gate", *Applied Physics Letters*, vol. 92, pp. 174106, 2008.

[DUR 08b] DURAND C. *et al.*, "Characterization of in-IC integrable in-plane nanometer scale resonators fabricated by a silicon on nothing advanced CMOS technology", *Proceedings of the IEEE MEMS conference*, pp. 1016–1019, 2008.

[DUR 08c] DURAND C. *et al.*, "In-plane silicon-on-nothing nanometer-scale resonant suspended gate MOSFET for in-IC integration perspectives", *IEEE Electron Device Letters*, vol. 29, pp. 494–496, 2008.

[EIC 07] EICHENFIELD M., MICHAEL C.P., PERAHIA R. *et al.*, "Actuation of micro-optomechanical systems via cavity-enhanced optical dipole forces", *Nature Photonics*, vol. 1, pp. 416–422, 2007.

[EIC 09] EICHENFIELD M., CAMACHO R., CHAN J. *et al.*, "A picogram- and nanometre-scale photonic-crystal optomechanical cavity", *Nature*, vol. 459, pp. 550–555, 2009.

[EKI 02] EKINCI K.L., YANG Y.T., HUANG X.M.H. *et al.*, "Balanced electronic detection of displacement in nanoelectromechanical systems", *Applied Physics Letters*, vol. 81, pp. 2253–2255, 2002.

[EKI 04a] EKINCI K.L., HUANG X.M.H., ROUKES M.L., "Ultrasensitive nanoelectromechanical mass detection", *Applied Physics Letters*, vol. 84, pp. 4469–4471, 2004.

[EKI 04b] EKINCI K.L., YANG Y.T., ROUKES M.L. *et al.*, "Ultimate limits to inertial mass sensing based upon nanoelectromechanical systems", *Journal of Applied Physics*, vol. 95, pp. 2682–2689, 2004.

[EOM 11] EOM K., PARK H.S., YOON D.S. *et al.*, "Nanomechanical resonators and their applications in biological/chemical detection: nanomechanics principles", *Physics Reports*, vol. 503, pp. 115–163, 2011.

[ERB 01] ERBE A., WEISS C., ZWERGER W. *et al.*, "Nanomechanical resonator shuttling single electrons at radio frequencies", *Physical Review Letters*, vol. 87, pp. 096106-1–096106-4, 2001.

[ERN 09] ERNST T. *et al.*, A 3D stacked nanowire technology – applications in advanced CMOS and beyond, International Conference on Frontiers of Characterization and Metrology for Nanoelectronics, 2009.

[ETA 08] ETAKI S. *et al.*, "Motion detection of a micromechanical resonator embedded in a d.c. SQUID", *Nature Physics*, vol. 4, pp. 785–788, 2008.

[ETT 14] ETTELT D. *et al.*, "3D magnetic field sensor concept for use in inertial measurement units (IMUs)", *Journal of Microelectromechanical Systems*, vol. 23, pp. 324–333, 2014.

[FAN 11] FANGET S. *et al.*, "Gas sensors based on gravimetric detection – a review", *Sensors and Actuators B: Chemical*, vol. 160, pp. 804–821, 2011.

[FED 08] FEDDER G.K., HOWE R.T., LIU T.K. *et al.*, "Technologies for cofabricating MEMS and electronics", *Proceedings of the IEEE*, vol. 96, pp. 306–322, 2008.

[FEN 07] FENG X.L., HE R., YANG P. *et al.*, "Very high frequency silicon nanowire electromechanical resonators", *Nano Letters*, vol. 7, pp. 1953–1959, 2007.

[FER 13] FERNANDEZ-REGULEZ M. *et al.*, "Horizontally patterned Si nanowire growth for nanomechanical devices", *Nanotechnology*, vol. 24, 095303, 2013.

[FON 02] FON W., SCHWAB K.C., WORLOCK J.M. *et al.*, "Phonon scattering mechanisms in suspended nanostructures from 4 to 40 K", *Physical Review B*, vol. 66, pp. 045302-1–045302-5, 2002.

[GAD 99] GAD-EL-HAK M., "The fluid mechanics of microdevices – the Freeman Scholar Lecture", *Journal of Fluids Engineering*, vol. 121, pp. 5–33, 1999.

[GIG 06] GIGAN S. *et al.*, "Self-cooling of a micromirror by radiation pressure", *Nature*, vol. 444, pp. 67–70, 2006.

[GOR 98] GORELIK L.Y. *et al.*, "Shuttle mechanism for charge transfer in Coulomb blockade nanostructures", *Physical Review Letters*, vol. 80, pp. 4526–4529, 1998.

[GOU 10] GOUTTENOIRE V. *et al.*, "Digital and FM demodulation of a doubly clamped single-walled carbon-nanotube oscillator: towards a nanotube cell phone", *Small*, vol. 6, pp. 1060–1065, 2010.

[GRO 04] GROB R.L., BARRY E.F., *Modern Practice of Gas Chromatography*, Wiley-Interscience, New York, 2004.

[GRO 08a] GROGG D., TEKIN H.C., MAZZA M. *et al.*, "Laterally vibrating-body double gate MOSFET with improved signal detection references", *Proceedings of the Device Research Conference*, pp. 155–156, 2008.

[GRO 08b] GROGG D. *et al.*, "Double gate movable body micro-electro-mechanical FET as hysteretic switch: application to data transmission systems", *Proceedings of the IEEE IEDM*, pp. 302–305, 2008.

[GRÖ 09] GRÖBLACHER S. *et al.*, "Demonstration of an ultracold micro-optomechanical oscillator in a cryogenic cavity", *Nature Physics*, vol. 5, pp. 485–488, 2009.

[GRY 10] GRYNBERG G., ASPECT A., FABRE C., *Introduction to Quantum Optics from the Semi-classical Approach to Quantized Light*, Cambridge University Press, 2010.

[GUO 04] GUO J., ZHAO Y., "Influence of Van der Waals and Casimir forces on electrostatic torsional actuators", *Journal of Microelectromechanical Systems*, vol. 13, pp. 1027–1035, 2004.

[HAL 13] HALL H.J., RAHAFROOZ A., BROWN J.J. *et al.*, "I-shaped thermally actuated VHF resonators with submicron components", *Sensors and Actuators A: Physical*, vol. 195, pp. 160–166, 2013.

[HAN 12] HANAY M.S. *et al.*, "Single-protein nanomechanical mass spectrometry in real time", *Nature Nanotechnology*, vol. 7, pp. 602–608, 2012.

[HAR 00] HARLEY J.A., KENNY T.W., "1/f noise considerations for the design and process optimization of piezoresistive cantilevers", *Journal of Microelectromechanical Systems*, vol. 9, pp. 226–235, 2000.

[HE 06] HE R., YANG P., "Giant piezoresistance effect in silicon nanowires", *Nature Nanotechnology*, vol. 1, pp. 42–46, 2006.

[HE 08] HE R., FENG X.L., ROUKES M.L. *et al.*, "Self-transducing silicon nanowire electromechanical systems at room", *Nano Letters*, vol. 8, pp. 1756–1761, 2008.

[HEC 08] HECK A.J.R., "Native mass spectrometry: a bridge between interactomics and structural biology", *Nature Methods*, vol. 5, pp. 927–933, 2008.

[HEN 04] HENKEL C., JOULAIN K., MULET J.-P. *et al.*, "Coupled surface polaritons and the Casimir force", *Physical Review A*, vol. 69, p. 023808, 2004.

[HEN 07] HENTZ S. *et al.*, "Importance of non-linearities and of quantum forces in NEMS design: modeling and experimental comparison", *Proceedings of the APCOM*, 2007.

[HER 54] HERRING C., VOGT E., "Transport and deformation-potential theory for many-valley semiconductors with anisotropic scattering", *Physical Review*, vol. 101, pp. 944–961, 1954.

[HER 09] HERTZBERG J.B. *et al.*, "Back-action-evading measurements of nanomechanical motion", *Nature Physics*, vol. 6, pp. 213–217, 2009.

[HOC 08] HOCHBAUM A.I. *et al.*, "Enhanced thermoelectric performance of rough silicon nanowires", *Nature*, vol. 451, pp. 163–168, 2008.

[HON 08] HONG K., KIM J., LEE S. *et al.*, "Strain-driven electronic band structure modulation of Si nanowires", *Nano Letters*, vol. 8, pp. 1335–1341, 2008.

[HOS 10] HOSSEIN-ZADEH M., VAHALA K.J., MEMBER S., "An optomechanical oscillator on a silicon chip", *IEEE Journal of Selected Topics in Quantum Electronics*, vol. 16, pp. 276–287, 2010.

[HSU 02] HSUEH C.-H., "Modeling of elastic deformation of multilayers due to residual stresses and external bending", *Journal of Applied Physics*, vol. 91, pp. 9652–9656, 2002.

[HU 05] HU Q., NOLL R.J., LI H. *et al.*, "The orbitrap: a new mass spectrometer", *Journal of Mass Spectrometry*, vol. 40, pp. 430–443, 2005.

[HUA 08] HUANG W. *et al.*, "Fully monolithic CMOS nickel microresonator oscillator", *Proceedings of the IEEE MEMS Conference*, pp. 3–6, 2008.

[HUS 03] HUSAIN A. *et al.*, "Nanowire-based very-high-frequency electromechanical resonator", *Applied Physics Letters*, vol. 83, pp. 1240–1242, 2003.

[ILC 06] ILCHENKO V.S., MATSKO A.B., "Optical resonators with whispering-gallery modes – part II: applications", *IEEE Journal of Quantum Electronics*, vol. 12, pp. 15–32, 2006.

[ILI 04] ILIC B., YANG Y., CRAIGHEAD H.G., "Virus detection using nanoelectromechanical devices", *Applied Physics Letters*, vol. 85, pp. 2604–2606, 2004.

[INT 13] INTRAVAIA F. *et al.*, "Strong Casimir force reduction through metallic surface nanostructuring", *Nature Communications*, vol. 4, p. 2515, 2013.

[IVA 11] IVALDI P. *et al.*, "50 nm thick AlN film-based piezoelectric cantilevers for gravimetric detection", *Journal of Micromechanics and Microengineering*, vol. 21, 085023, 2011.

[JEN 07] JENSEN K., WELDON J., GARCIA H. *et al.*, "Nanotube radio", *Nano Letters*, vol. 7, pp. 3508–3511, 2007.

[JOU 09] JOURDAN G., LAMBRECHT A., COMIN F. *et al.*, "Quantitative non-contact dynamic Casimir force measurements", *Europhysics Letters*, vol. 85, p. 31001, 2009.

[JOU 13] JOURDAN G. *et al.*, "NEMS-based heterodyne self-oscillator", *Sensors and Actuators A: Physical*, vol. 189, pp. 512–518, 2013.

[JU 05] JU Y.S., "Phonon heat transport in silicon nanostructures", *Applied Physics Letters*, vol. 87, p. 153106, 2005.

[KAC 09] KACEM N., HENTZ S., PINTO D. *et al.*, "Nonlinear dynamics of nanomechanical beam resonators: improving the performance of NEMS-based sensors", *Nanotechnology*, vol. 20, 275501, 2009.

[KAC 10] KACEM N., ARCAMONE J., PEREZ-MURANO F. *et al.*, "Dynamic range enhancement of nonlinear nanomechanical resonant cantilevers for highly sensitive NEMS gas/mass sensor", *Journal of Micromechanics and Microengineering*,vol. 20, 045023, 2010.

[KAN 82] KANDA Y., "A graphical representation of the piezoresistance coefficients in silicon", *IEEE Transactions on Electron Devices*, vol. 29, pp. 64–70, 1982.

[KAR 09] KARABALIN R.B. *et al.*, "Piezoelectric nano electromechanical resonators based on aluminum nitride thin films", *Applied Physics Letters*, vol. 95, 103111, 2009.

[KIM 12] KIM K.H., KIM B.H., SEO Y.H., "A noncontact intraocular pressure measurement device using a micro reflected air pressure sensor for the prediagnosis of glaucoma", *Journal of Micromechanics and Microengineering*, vol. 22, 035022, 2012.

[KIP 08] KIPPENBERG T.J., VAHALA K.J., "Cavity optomechanics: back-action at the mesoscale", *Science*, vol. 321, pp. 1172–1176, 2008.

[KIT 98] KITTEL C., *Physique de l'état solide*, Dunod, Paris, p. 610, 1998.

[KLI 09] KLIMCHITSKAYA G.L., MOSTEPANENKO V.M., "The Casimir force between real materials: experiment and theory", *Reviews of Modern Physics*, vol. 81, pp. 1827–1885, 2009.

[KNO 03] KNOBEL R.G., CLELAND A.N., "Nanometre-scale displacement sensing using a single electron transistor", *Nature*, vol. 424, pp. 291–293, 2003.

[KOU 11] KOUMELA A. *et al.*, "Piezoresistance of top-down suspended Si nanowires", *Nanotechnology*, vol. 22, 395701, 2011.

[KOU 13] KOUMELA A. *et al.*, "High frequency top-down junction-less silicon nanowire resonators", *Nanotechnology*, vol. 24, 435203, 2013.

[LAH 04] LAHAYE M.D., BUU O., CAMAROTA B. *et al.*, "Approaching the quantum limit of a nanomechanical resonator", *Science*, vol. 304, pp. 74–77, 2004.

[LAH 09] LAHAYE M.D., SUH J., ECHTERNACH P.M. *et al.*, "Nanomechanical measurements of a superconducting qubit", *Nature*, vol. 459, pp. 960–964, 2009.

[LAH 13] LAHTEENMAKI P., PARAOANU G.S., HASSEL J., "Dynamical Casimir effect in a Josephson metamaterial", *Proceedings of the National Academy of Sciences*, vol. 110, pp. 4234–4238, 2013.

[LAM 97] LAMOREAUX S.K., "Demonstration of the Casimir force in the 0.6 to 6 mm range", *Physical Review Letters*, vol. 78, pp. 5–8, 1997.

[LAM 07a] LAMBRECHT A., PIROZHENKO I., DURAFFOURG L. *et al.*, "The Casimir effect for silicon and gold slabs", *Europhysics Letters*, vol. 77, p. 44006, 2007.

[LAM 07b] LAMOREAUX S.K., "Casimir forces: still surprising after 60 years", *Physics Today*, vol. 60, pp. 40–45, 2007.

[LEE 08] LEE H.J., PARK K.K., ORALKAN O. *et al.*, "CMUT as a chemical sensor for DMMP detection", *IEEE International Frequency Control Symposium*, pp. 434–439, 2008.

[LEU 04] LEUS V., ELATA D., Fringing field effect in electrostatic actuators, Technical report ETR-2004-2 from TECHNION – Israel Institute of Technology, Faculty of Mechanical Engineering, 2004.

[LI 03] LI D. *et al.*, "Thermal conductivity of individual silicon nanowires", *Applied Physics Letters*, vol. 83, pp. 2934–2936, 2003.

[LI 07] LI M.O., TANG H.X., ROUKES M.L., "Ultra-sensitive NEMS-based cantilevers for sensing, scanned probe and very high-frequency applications", *Nature Nanotechnology*, vol. 2, pp. 114–120, 2007.

[LI 08] LI M., PERNICE W.H.P., XIONG C. *et al.*, "Harnessing optical forces in integrated photonic circuits", *Nature*, vol. 456, pp. 480–485, 2008.

[LI 13] LI S., "Advances of CMOS-MEMS technology for resonator applications", *Proceedings of the IEEE NEMS Conference*, vol. 1, pp. 5–8, 2013.

[LIF 08] LIFSHITZ R., CROSS M.C., *Reviews of Nonlinear Dynamics and Complexity*, in SCHUSTER H.G. (ed.), Wiley, New York, 2008.

[LIN 09] LIN Q., ROSENBERG J., JIANG X. *et al.*, "Mechanical oscillation and cooling actuated by the optical gradient force", *Physical Review Letters*, vol. 103, p. 103601, 2009.

[LIN 10] LIN Q. *et al.*, "Coherent mixing of mechanical excitations in nano-optomechanical structures", *Nature Photonics*, vol. 4, pp. 236–242, 2010.

[LIU 05] LIU X. *et al.*, "A loss mechanism study of a very high Q silicon micromechanical oscillator", *Journal of Applied Physics*, vol. 97, p. 023524, 2005.

[LOH 12] LOH O.Y., ESPINOSA H.D., "Nanoelectromechanical contact switches", *Nature Nanotechnology*, vol. 7, pp. 283–295, 2012.

[LOP 09] LOPEZ J.L. *et al.*, "Integration of RF-MEMS resonators on submicrometric commercial CMOS technologies", *Journal of Micromechanics and Microengineering*, vol. 19, 015002, 2009.

[LUN 04] LUND E., FINSTAD T.G., "Design and construction of a four-point bending based set-up for measurement of piezoresistance in semiconductors", *Review of Scientific Instruments*, vol. 75, p. 4960, 2004.

[MA 12] MA J., POVINELLI M.L., "Applications of optomechanical effects for on-chip manipulation of light signals", *Current Opinion in Solid State and Materials Science*, vol. 16, pp. 82–90, 2012.

[MAD 11] MADOU M., *Fundamentals of Microfabrication*, CRC Press, Boca Raton, Flonda, USA, 2011.

[MAH 12] MAHBOOB I., NISHIGUCHI K., OKAMOTO H. *et al.*, "Phonon-cavity electromechanics", *Nature Physics*, vol. 8, pp. 387–392, 2012.

[MAK 00] MAKAROV A., "Electrostatic axially harmonic orbital trapping: a high-performance technique of mass analysis", *Analytical Chemistry*, vol. 72, pp. 1156–1162, 2000.

[MAR 12] MARATHE R., WANG W., WEINSTEIN D., "Si-based unreleased hybrid MEMS-CMOS resonators in 32 nm technology", *Proceedings of the IEEE MEMS conference*, pp. 1–4, 2012.

[MAR 14] MARATHE R. *et al.*, "Resonant body transistors in IBM's 32 nm", *Journal of Microelectromechanical Systems*, vol. 23, pp. 636–650, 2014.

[MAT 06] MATSKO A.B., ILCHENKO V.S., "Optical resonators with whispering-gallery modes – part I: basics", *IEEE Journal of Quantum Electronics*, vol. 12, pp. 3–14, 2006.

[MET 08] METZGER C., FAVERO I., ORTLIEB A. *et al.*, "Optical self cooling of a deformable Fabry-Perot cavity in the classical limit", *Physical Review B*, vol. 78, 035309, 2008.

[MID 00] MIDZOR M.M. *et al.*, "Imaging mechanisms of force detected FMR microscopy", *Journal of Applied Physics*, vol. 87, p. 6493, 2000.

[MIL 10] MILE E. *et al.*, "In-plane nanoelectromechanical resonators based on silicon nanowire piezoresistive detection", *Nanotechnology*, vol. 21, 165504, 2010.

[MUN 09] MUNDAY J.N., CAPASSO F., PARSEGIAN V.A., "Measured long-range repulsive Casimir-Lifshitz forces", *Nature*, vol. 457, pp. 170–173, 2009.

[MUÑ 13] MUÑOZ-GAMARRA J.L. *et al.*, "Integration of NEMS resonators in a 65nm CMOS technology", *Microelectronic Engineering*, vol. 110, pp. 246–249, 2013.

[NAI 06] NAIK A.K., *et al.*, "Cooling a nanomechanical resonator with quantum back-action", *Nature*, vol. 443, pp. 193–196, 2006.

[NAI 09] NAIK A.K., HANAY M.S., HIEBERT W.K. *et al.*, "Towards single-molecule nanomechanical mass spectrometry", *Nature Nanotechnology*, vol. 4, pp. 445–450, 2009.

[NAK 90] NISHIYAMA H., NAKAMURA M., "Capacitance of a strip capacitor" *IEEE Transactions on Components Hybrids and Manufacturing Technology*, vol. 13, pp. 417–423, 1990.

[NAK 99] NAKAMURA Y., PASHKIN Y.A., TSAI J.S., "Coherent control of macroscopic quantum states in a single-Cooper-pair box", *Nature*, vol. 398, pp. 786–788, 1999.

[NAT 67] NATHANSON H.C., NEWELL W.E., WICKSTROM R.A. *et al.*, "The resonant gate transistor", *IEEE Transactions on Electron Devices*, vol. ED-14, pp. 117–133, 1967.

[NAY 04] NAYFEH A.H., *Applied Nonlinear Dynamics and Experimental Methods*, Wiley-VCH Verlag GmbH, Berlin, 2004.

[NET 05a] NETO P.A.M., LAMBRECHT A., REYNAUD S., "Casimir effect with rough metallic mirrors", *Physical Review A*, vol. 72, 012115, 2005.

[NET 05b] NETO P.A.M., LAMBRECHT A., REYNAUD S., "Roughness correction to the Casimir force: beyond the proximity force approximation", *Europhysics Letters (EPL)*, vol. 69, pp. 924–930, 2005.

[NGU 99] NGUYEN C.T.-C., HOWE R.T., "An integrated CMOS micromechanical resonator high-Q oscillator", *IEEE Journal of Solid-State Circuits*, vol. 34, pp. 440–455, 1999.

[NGU 00] NGUYEN C.T., "Micromechanical circuits for communication transceivers", *Proceedings of the Bipolar/BiCMOS Circuits and Technology Meeting*, pp. 142–149, 2000.

[O'CO 10] O'CONNELL A.D. *et al.*, "Quantum ground state and single-phonon control of a mechanical resonator", *Nature*, vol. 464, pp. 697–703, 2010.

[OLL 12] OLLIER E. *et al.*, "Ultra-scaled high-frequency single-crystal Si NEMS resonators and their front-end co-integration with CMOS for high sensitivity applications", *Proceedings of the IEEE MEMS conference*, pp. 1368–1371, 2012.

[PAL 85] PALIK E.D., *Handbook of Optical Constants of Solids*, Academic Press, p. 3187, 1985.

[PAL 05] PALASANTZAS G., DE HOSSON J.T.M., "Pull-in characteristics of electromechanical switches in the presence of Casimir forces: influence of self-affine surface roughness", *Physical Review A*, vol. 72, p. 115426, 2005.

[PAP 14] PAPPAKRISHNAN V.K., MUNDRU P.C., GENOV D.A., "Repulsive Casimir force in magnetodielectric plate configurations", *Physical Review B*, vol. 89, p. 045430, 2014.

[PAR 03] PARKIN S. *et al.*, "Magnetically engineered spintronic sensors and memory", *Proceedings of the IEEE*, vol. 91, pp. 661–680, 2003.

[PHI 14] PHILIPPE J. *et al.*, "Fully monolithic and ultra-compact NEMS-CMOS self-oscillator based on single-crystal silicon resonators and low-cost CMOS circuitry", *Proceedings of the IEEE MEMS Conference*, pp. 1071–1074, 2014.

[PIN 00] PINARD M., COHADON P.F., BRIANT T. *et al.*, "Full mechanical characterization of a cold damped mirror", *Physical Review E*, vol. 63, p. 013808, 2000.

[POO 11] POOT M., ETAKI S., YAMAGUCHI H. *et al.*, "Discrete-time quadrature feedback cooling of a radio-frequency mechanical resonator", *Applied Physics Letters*, vol. 99, p. 013113, 2011.

[POS 05] POSTMA H.W.C., KOZINSKY I., HUSAIN A. *et al.*, "Dynamic range of nanotube- and nanowire-based electromechanical systems", *Applied Physics Letters*, vol. 86, p. 223105, 2005.

[POU 03] POURKAMALI S. *et al.*, "High-Q single crystal silicon HARPSS capacitive beam resonators with self-sligned sub-100-nm transduction gaps", *Journal of Microelectromechanical Systems*, vol. 12, pp. 487–496, 2003.

[POV 05] POVINELLI M.L., SMYTHE E.J., JOHNSON S.G. *et al.*, "Evanescent-wave bonding between optical waveguides", *Optics Letters*, vol. 30, pp. 3042–3044, 2005.

[RAH 11] RAHAFROOZ A., POURKAMALI S., "High frequency dual-mode thermal-piezoresistive oscillators", *Proceedings of the IFCS*, pp. 1–4, 2011.

[RAZ 96] RAZAVI B., "A study of phase noise in CMOS oscillators", *IEEE Journal of Solid-State Circuits*, vol. 31, pp. 331–343, 1996.

[REY 02] REYNAUD S., The Casimir force and the quantum theory of optical networks, Lecture at the Physics Graduate School of ENS (Paris, France), p. 100, 2002.

[RIC 08] RICHTER J., PEDERSEN J., BRANDBYGE M. *et al.*, "Piezoresistance in p-type silicon revisited", *Journal of Applied Physics*, vol. 104, p. 023715, 2008.

[ROB 82] ROBINS W.P., *Phase Noise in Signal Sources*, Peter Peregrinus Ltd., 1982.

[ROB 83] ROBINS W.P., *Phase Noise in Signal Sources*, The Institution of Engineering and Technology, 1983.

[ROB 09] ROBERT P. *et al.*, "M&NEMS: a new approach for ultra-low cost 3D inertial sensor", *Proceedings of the IEEE Sensors (IEEE)*, pp. 963–966, 2009.

[ROD 11] RODRIGUEZ A.W., CAPASSO F., JOHNSON S.G., "The Casimir effect in microstructured geometries", *Nature Photonics*, vol. 5, pp. 211–221, 2011.

[ROW 05] ROWE D.M., *Thermoelectrics Handbook: Macro to Nano*, CRC Press, 2005.

[ROW 08] ROWE A.C.H., "Silicon nanowires feel the pinch", *Nature Nanotechnology*, vol. 3, pp. 311–312, 2008.

[RUB 05] RUBIOLA E., "On the measurement of frequency and of its sample variance with high-resolution counters", *Review of Scientific Instruments*, vol. 76, 054703, 2005.

[RUG 04] RUGAR D., BUDAKIAN R., MAMIN H.J. *et al.*, "Single spin detection by magnetic resonance force microscopy", *Nature*, vol. 430, pp. 329–332, 2004.

[SAF 13] SAFAVI-NAEINI A.H. *et al.*, "Squeezed light from a silicon micromechanical resonator", *Nature*, vol. 500, pp. 185–189, 2013.

[SAG 13] SAGE E. *et al.*, "Frequency-addressed NEMS arrays for mass and gas sensing applications", *Proceedings of the Transducers Conference*, pp. 665–668, 2013.

[SAU 77] SAUVAGE G., "Phase noise in oscillators: a mathematical analysis of Leeson's model", *Transaction on Instrumentation and Measurement*, vol. 26, pp. 408–410, 1977.

[SAZ 04] SAZONOVA V. *et al.*, "A tunable carbon nanotube electromechanical oscillator", *Nature*, vol. 431, pp. 284–287, 2004.

[SCH 05] SCHWAB K.C., ROUKES M.L., "Putting mechanics into nanoelectromechanical structures are starting to approach the ultimate quantum", *Physics Today*, vol. 58, pp. 36–42, 2005.

[SCH 00a] SCHOUTEN R.N., HARMANS C.J.P.M., ORLANDO T.P., "Quantum superposition of macroscopic persistent-current states", *Science*, vol. 290, pp. 773–777, 2000.

[SCH 00b] SCHWAB K., HENRIKSEN E., WORLOCK J. *et al.*, "Measurement of the quantum of thermal conductance", *Nature*, vol. 404, pp. 974–977, 2000.

[SCH 06] SCHLIESSER A., DEL'HAYE P., NOOSHI N. *et al.*, "Radiation pressure cooling of a micromechanical oscillator using dynamical backaction", *Physical Review Letters*, vol. 97, 243905, 2006.

[SCH 08] SCHLIESSER A., RIVIÈRE R., ANETSBERGER G. *et al.*, "Resolved-sideband cooling of a micromechanical oscillator", *Nature Photonics*, vol. 4, pp. 415–419, 2008.

[SCH 09a] SCHLIESSER A., ARCIZET O., RIVIÈRE R. *et al.*, "Resolved-sideband cooling and position measurement of a micromechanical oscillator close to the Heisenberg uncertainty limit", *Nature Physics*, vol. 5, pp. 509–514, 2009.

[SCH 09b] SCHMIDT V., WITTEMANN J.V., SENZ S. *et al.*, "Silicon nanowires: a review on aspects of their growth and their electrical properties", *Advanced Materials*, vol. 21, pp. 2681–2702, 2009.

[SIR 09] SIRIA A. *et al.*, "Viscous cavity damping of a microlever in a simple fluid", *Physical Review Letters*, vol. 102, p. 254503, 2009.

[SIR 10] SIRIA A., Systèmes nano eléctro mécanique et interactions à l'échelle nanométrique, PhD thesis, Joseph-Fourier University, Grenoble, France, 2010.

[SNY 08] SNYDER G.J., TOBERER E.S., "Complex thermoelectric materials", *Nature Materials*, vol. 7, pp. 105–114, 2008.

[SRI 11] SRINIVASAN K., MIAO H., RAKHER M.T. *et al.*, "Optomechanical transduction of an integrated silicon cantilever probe using a microdisk resonator", *Nano Letters*, vol. 11, pp. 791–797, 2011.

[SZE 81] SZE S.M., *Physics of Semiconductor Devices*, Wiley-Interscience, New York, vol. 867, 1981.

[TAN 02] TANG H.X., HUANG X.M.H., ROUKES M.L. *et al.*, "Two-dimensional electron-gas actuation and transduction for GaAs nanoelectromechanical systems", *Applied Physics Letters*, vol. 81, pp. 3879–3881, 2002.

[THI 00] THIELICKE E., OBERMEIER E., "Microactuators and their technologies", *Mechatronics*, vol. 10, pp. 431–455, 2000.

[THO 10] THOURHOUT D.V., ROELS J., "Optical gradient force", *Nature*, vol. 4, pp. 211–217, 2010.

[TIM 25] TIMOSHENKO B.Y.S., "Analysis of bi-metal thermostats", *Journal of the Optical Society of America*, vol. 11, pp. 233–255, 1925.

[TRU 07] TRUITT P.A., HERTZBERG J.B., HUANG C.C. *et al.*, "Efficient and sensitive capacitive readout of nanomechanical resonator arrays", *Nano Letters*, vol. 7, pp. 120–126, 2007.

[TSA 12] TSAI M., LIU Y., FANG W., "A three-axis CMOS-MEMS accelerometer structure with vertically integrated fully differential sensing electrodes", *Journal of Microelectromechanical Systems*, vol. 21, pp. 1329–1337, 2012.

[UET 08] UETRECHT C. *et al.*, "High-resolution mass spectrometry of viral assemblies: molecular composition and stability of dimorphic hepatitis B virus capsids", *Proceedings of the National Academy of Sciences of the United States of America*, vol. 105, pp. 9216–9220, 2008.

[VAH 03] VAHALA K.J., "Optical microcavities", *Nature*, vol. 424, pp. 839–846, 2003.

[VAN 10] VAN THOURHOUT D., ROELS J., "Optical gradient force", *Nature Photonics*, vol. 4, pp. 211–217, 2010.

[VEI 01] VEIJOLA T., TUROWSKI M., "Compact damping models for laterally moving microstructures with gas-rarefaction effects", *Journal of Microelectromechanical Systems*, vol. 10, pp. 263–273, 2001.

[VER 06a] VERD J. *et al.*, "Integrated CMOS – MEMS with on-chip readout electronics for high-frequency applications", *IEEE Electron Device Letters*, vol. 27, pp. 495–497, 2006.

[VER 06b] VERNOTTE F., Techniques de l'Ingénieur, R 681-1-10, 2006.

[VER 07a] VERD J., URANGA A., ABADAL G. *et al.*, "Monolithic mass sensor fabricated using a conventional technology with attogram resolution in air conditions", *Applied Physics Letters*, vol. 91, p. 013501, 2007.

[VER 07b] VERD J. *et al.*, "CMOS cantilever-based oscillator for attogram mass sensing", *IEEE International Symposium on Circuits and Systems*, IEEE, pp. 3319–3322, 2007.

[VER 08] VERD J. *et al.*, Monolithic CMOS MEMS oscillator circuit for sensing in the attogram range", *IEEE Electron Device Letters*, vol. 29, pp. 146–148, 2008.

[VER 13] VERD J., URANGA A., SEGURA J. *et al.*, "A 3V CMOS-MEMS oscillator in 0.35 μm CMOS technology", *Proceedings of the Transducers Conference*, pp. 806–809, 2013.

[VIO 02] VION D. *et al.*, "Manipulating the quantum state of an electrical circuit", *Science*, vol. 296, pp. 886–889, 2002.

[WAL 12] WALTHER A. *et al.*, "3-axis gyroscope with Si nanogage piezo-resistive detection", *Proceedings of the IEEE MEMS conference*, pp. 480–483, 2012.

[WAN 03] WANG Y., HENRY J.A., ZEHNDER A.T. *et al.*, "Surface chemical control of mechanical energy losses in micromachined silicon structures", *Journal of Physical Chemistry*, vol. 107, pp. 14270–14277, 2003.

[WEA 90] WEAVER W., TIMOSHENKO S.P., YOUNG D.H., *Vibration Problems in Engineering*, Wiley, New York, 1990.

[YAN 00] YANG J., ONO T., ESASHI M., "Mechanical behavior of ultrathin microcantilever", *Sensors & Actuators: A. Physical*, vol. 82, pp. 102–107, 2000.

[YAN 06] YANG Y.T., CALLEGARI C., FENG X.L. *et al.*, "Zeptogram-scale nanomechanical mass sensing", *Nano Letters*, vol. 6, pp. 583–586, 2006.

[YAZ 98] YAZDI N., AYAZI F., NAJAFI K., "Micromachined inertial sensors", *Proceedings of the IEEE*, vol. 86, pp. 1640–1659, 1998.

[YOL 12] YOLE DÉVELOPPEMENT, Technology Trends for Inertial MEMS, pp. 1–12, available at: www.yole.fr, 2012.

[YON 88] YONG Y.K., VIG J.R., "Resonator surface contamination – a cause of frequency fluctuation?", *Proceedings of the 42nd Annual Frequency Control Symposium*, pp. 397–403, 1988.

[ZAL 10] ZALALUTDINOV M.K. *et al.*, "CMOS-integrated RF MEMS resonators", *Journal of Microelectromechanical Systems*, vol. 19, pp. 807–815, 2010.

[ZEN 38] ZENER C., "Internal friction in solids II. General theory of thermoelastic internal friction", *Physical Review*, vol. 53, pp. 90–99, 1938.

[ZHA 09] ZHAO R., ZHOU J., KOSCHNY T. *et al.*, "Repulsive Casimir force in chiral metamaterials", *Physical Review Letters*, vol. 103, p. 103602, 2009.

[ZHA 12] ZHANG W.-M., MENG G., WEI X., "A review on slip models for gas microflows", *Microfluidics and Nanofluidics*, vol. 13, pp. 845–882, 2012.

Index

S, T

Other titles from

ISTE

in

Electronics Engineering

2014

CONSONNI Vincent, FEUILLET Guy
Wide Band Gap Semiconductor Nanowires 1: Low-Dimensionality Effects and Growth
Wide Band Gap Semiconductor Nanowires 2: Heterostructures and Optoelectronic Devices

GAUTIER Jean-Luc
Design of Microwave Active Devices

LACAZE Pierre Camille, LACROIX Jean-Christophe
Non-volatile Memories

TEMPLIER François
OLED Microdisplays: Technology and Applications

2013

COSTA François, GAUTIER Cyrille, LABOURE Eric, REVOL Bertrand
Electromagnetic Compatibility in Power Electronics

KORDON Fabrice, HUGUES Jérôme, CANALS Agusti, DOHET Alain
Embedded Systems: Analysis and Modeling with SysML, UML and AADL

LE TIEC Yannick
Chemistry in Microelectronics

2012

BECHERRAWY Tamer
Electromagnetism: Maxwell Equations, Wave Propagation and Emission

SAGUET Pierre
Numerical Analysis in Electromagnetics: The TLM Method

2011

ALGANI Catherine, RUMELHARD Christian, BILLABERT Anne-Laure
Microwaves Photonic Links: Components and Circuits

BAUDRANT Annie
Silicon Technologies: Ion Implantation and Thermal Treatment

DEFAY Emmanuel
Integration of Ferroelectric and Piezoelectric Thin Films: Concepts ans Applications for Microsystems

DEFAY Emmanuel
Ferroelectric Dielectrics Integrated on Silicon

BESNIER Philippe, DÉMOULIN Bernard
Electromagnetic Reverberation Chambers

LANDIS Stefan
Nano-lithography

2010

LANDIS Stefan
Lithography

PIETTE Bernard
VHF / UHF Filters and Multicouplers

2009

DE SALVO Barbara
Silicon Non-volatile Memories / Paths of Innovation

DECOSTER Didier, HARARI Joseph
Optoelectronic Sensors

FABRY Pierre, FOULETIER Jacques
Chemical and Biological Microsensors / Applications in Fluid Media

GAUTIER Jacques
Physics and Operation of Silicon Devices in Integrated Circuits

MOLITON André
Solid-State Physics for Electronics

PERRET Robert
Power Electronics Semiconductor Devices

SAGUET Pierre
Passive RF Integrated Circuits

2008

CHARRUAU Stéphane
Electromagnetism and Interconnections

Printed and bound by CPI Group (UK) Ltd, Croydon, CR0 4YY

13/09/2026

14961079-0001